Left in the Dark

(The Biological Origins of the Fall From Grace)

An investigation into the evolution of the human brain.

A journey to the edge of the human mind.

By

Graham Gynn and Tony Wright

Left in the Dark

First edition Lulu Books 2007
Second revised edition Kaleidos Press 2008
ISBN 978-0-9556784-0-0

Cover S. Gynn & T. Wright
Line illustrations L. Clifford
Diagrams S. Charter & T. Wright

www.leftinthedark.org.uk

Acknowledgements and Dedications

I would like to dedicate my work here to the memory of Gordon Gynn - father, journalist and Cornishman.

Graham Gynn May 2007

It would take several pages to list everyone who has contributed in one way or another to the research that has resulted in the publication of this book. For this reason I will simply say thanks to all those who have helped over the last 16 years.

In loving memory of Grace (Granny) Rippingle.

For Corriene, Lexi and Jago.

Tony Wright May 2007

Legal Note

The theory and combined techniques outlined in this book are the subject of ongoing research and are not intended as medical advice. Deciding to adopt any of the approaches discussed may in some cases be detrimental; any such decision to do so is made entirely at your own discretion.

Contents

~ψ~

Foreword

~ψ~

The progress of science, and indeed, of human knowledge, requires a dynamic tension between the mere accumulation of observations and "dusty facts" and a synthetic process in which the accumulated results of scientific observation and inquiry are woven together into frameworks that, in the ideal case, create revolutionary paradigms that enhance human understanding of apparently discrete and unrelated aspects of nature. The hypotheses proposed in this book may well represent such a revolutionary paradigm. These ideas do not originate from the mainstream of academia, but rather are the contribution of two independent scholars. The history of science and intellectual inquiry teach us that, as is so often the case with truly novel syntheses, established scientific and intellectual institutions are too ossified, and too invested in the conventionally accepted worldview, to allow the introduction of a new paradigm without putting up considerable resistance.

Resistance will more than likely characterize the response to this book; its authors will undoubtedly be denounced as mavericks, unqualified to comment on such a momentous topic as the evolution of human consciousness; the ideas put forth here will be condemned as heresy. Indeed they **are** heresy, in the context of what we think we understand about human evolution, particularly the anomalous evolution of the human brain and consciousness. But one is reminded of the famous observation of philosopher Arthur Schopanhauer: All truth, he said, passes through three stages. First, it is ridiculed; second, it is violently opposed; third, it is accepted as being self-evident. We should be wary of rejecting out of hand the premises of a hypothesis that may one day seem self-evident.

Evolutionary biologists have long been puzzled by what is perhaps the chief mystery of human origins: the explosive and rapid expansion of the human brain in size and complexity over a vanishingly small span of evolutionary time. There is also the mystery of hemispheric lateralization and the apparent de-integration of the right- and left-hemispheric functions that we humans suffer. In this work, the authors postulate that it was not always so; the universal myth of a pre-historic Golden Age, they maintain, is a racial memory that reflects our primate evolution in an arboreal, rainforest environment in which humans possessed mental and psychic abilities that have since become lost or atrophied in the profane ages that followed. That rainforest environment favored a frugivorous diet rich in flavonoids, MAO inhibitors, and neurotransmitter precursors, and relatively low in steroid containing or inducing elements. This dietary regime both mimicked and fostered a state, reinforced by positive feedback loops, in which pineal functions, including neocortical expansion and hemispheric integration, were potentiated; moreover, these neurochemical feedback loops were amplified in succeeding generations via the regulation of gene expression in the developing foetus, independent of conventional evolutionary mechanisms of mutation and natural selection. Climate changes or other environmental catastrophes forced several lineages of hominids as well as archaic/early humans out of their forest-dwelling ancestral home into much harsher savannah or grassland environments. As a consequence dietary regimens shifted toward roots, tubers, grass seed and a greater proportion of animal protein, triggering a reversal of the positive feedback loops that had sustained pineal potentiation and hemispheric integration in the paradisiacal, forest-dwelling Golden Age. Pineal dominance was disrupted by steroid-mediated, testosterone-driven functions primarily due to the reduced consumption of flavonoids and other steroid-inhibitory dietary factors. Changes in the dietary patterns that were forced on the population by this migration put an end to the rapid evolution of the human brain and triggered its devolution, ultimately resulting in the damaged human neural architecture that we suffer from today, and the myriad mental and physical deficits that are the legacy of our biological 'fall from grace'.

It is not the place of a foreword to present the central tenets of a complex theory in detail; what is alluded to here is only the barest outline of an elegant hypothesis that plausibly elucidates many baffling aspects of human evolution, brain science, and physiology into a coherent explanatory framework. Ecologists have realized for several decades that the complex interrelations of plants and insects are largely mediated through plant chemistry, and that the interactive dynamics we can observe in these processes is a reflection of millions of years of plant-insect co-evolution. Evolutionary biologists have long suspected that similar co-evolutionary processes, mediated by interactions with plant secondary products, have influenced the evolution of vertebrates, including primates. The hypotheses presented in this book are incomplete, and are even now being refined and developed; however, even in their present form they present a credible foundation on which to build a better understanding of who we are, and how our puzzling human species got to be the way it is.

Dennis J. McKenna, Ph.D.
Burnaby, British Columbia, Canada
December 2007

Introduction

~ψ~

One evening, about half way though writing this book, I had a game of table tennis with my teenage daughter. We were both feeling a little stressed from our respective days. We wanted to talk and table tennis often helps the two-way communication. Something about hitting the ball to each other enhances communication. On this occasion it wasn't quite working so on a whim we both swapped hands. We began to play left-handed. The change was immediate and very apparent. We both became calm, tension dropped away and the peaceful atmosphere that ensued was quite palpable. We looked at each other and were amazed at the difference. I knew at that moment that Tony and I were on to something that was not a mere scientific or intellectual curiosity but something real and profound.

Tony has been working on the issue of brain organisation and perception for over fifteen years, and we have been collaborating since 2002. His role is inspiration and research, mine translation and scribe. The theory presented in this book has grown and developed during this time, and it is heartening that, every week it seems, research is appearing that not only fits but also supports this new paradigm. We are in contact with many of these researchers who have offered help and encouragement, and we are especially grateful to those who have added their comments here.

The pivotal moment for the thesis came in 1995 after Tony spent three days and nights awake. His over-stretched left hemisphere fell asleep leaving his right awake, functional and free of the left's influence. The next twenty minutes were not only euphoric but also deeply intriguing. During this brief window he investigated the capabilities of his unhindered right hemisphere and found its perceptual abilities superior to his normal self. This experiential research confirmed Tony's initial ideas of laterality and brain function, the implications of which are far-reaching. We could all be suffering from an evolutionary glitch that has affected how we perceive, think and behave.

If we look at our global society, it is apparent that all is not well. Despite good intentions and attempts at cooperation, we live in a very fragmented and violent world. There is war and genocide, we are inflicting havoc on the only planet that sustains us, and we are having increasing problems with interpersonal relationships. It seems we are incapable of behaving anywhere near the ideal we would like to maintain. These problems are becoming more intense in our present era as increasing population and dwindling resources exert more and more pressure.

The Golden Age

It was not always so. The earliest of times, according to the classical writer Hesiod, was a 'Golden Age'. Men lived as gods, with their hearts free from sorrow, in a land abundant in fruit and rich in flocks. The Greek philosopher Dicaearchus, of the late fourth century BC, tells us more – god-like men lived a life of leisure, health, peace and friendship, without care or toil, or the desires that lead to feuds and wars. 'Their life was easy for their food and all things grew spontaneously.' But these halcyon days came to an end – there was progressive degeneration through the ages of Silver, Brass, Heroes and Iron.

This step by step decline is a universal theme. The Hindu tradition identifies four epochs and each one was marked by a decline in moral and physical standards. The Kriti Yuga was the perfect age. Man had no worldly desires, disease, sorrows or fears. There was supreme happiness, continual delight and the ability to move about at will. 'The things the people needed spontaneously sprang from the earth everywhere and always whenever the mind desired it, and there was no need for houses either.' The Treta, Dvapara and Kali Yugas followed this wonderful Kriti Yuga. With each age, man's virtue lessened a quarter so in the Kali Yuga, our present age, only one quarter of man's virtue remains. Now, as is very apparent, we are afflicted by disease and suffering. Men have turned to wickedness, decadence and materialism. There is pain, sorrow, continual dissatisfaction and craving.

The Fall of Mankind

Nearly every culture has a tradition about this fall of mankind. In almost every one it came about because man strayed from the way of the gods. Adam disobeyed God's instructions and ate the forbidden fruit. In a Zambian myth, the first man, Kamonu, started to kill the animals created by the god Nyambi and was then forced from his garden. And the Hopi say that when the people began to depart from the instruction of the Great Spirit:

'There came among them a handsome one [...] in the form of a snake with a big head. He led the people still further away from one another and their pristine wisdom. They became suspicious of one another and accused one another wrongfully until they became fierce and warlike and began to fight one another.'

Richard Heinberg in 'Memories and Visions of Paradise' says: ' nearly every tradition ascribes the loss of Paradise to the appearance of some tragic aberration in the attitude or behaviour of human beings. While in the Golden Age they had been 'truth-speaking' and 'self-subdued', living with 'no evil desires, without guilt or crime', they now succumbed to suspicion, fear, greed, mistrust and violence.' Something happened which resulted in the loss of a former state of divine beingness. In its place arose the state of fear, mistrust and craving which has led to all the woes that we are experiencing today. Paradise has been transformed by our machinations into our present materialist, fear-based age of plastic and prozac.

Could these myths actually be a cultural remembering of a time when we really were perceptually more complete; to a time in which there was less fear, less violence, less craving and more contentment? If so, today, in an age characterised by so much distraction, we have truly forgotten what we were and who we are.

Two Perspectives

The mythic traditions of paradise allude to our naked, forest-dwelling, fruit-eating past. Various cataclysmic disasters portrayed in tales of floods, vulcanism and meteor impact brought the days of perfection to an end. These disruptive, earth-shattering events initiated a change in man too – a single divine self was split into two and the more fallen, delusional self assumed overall control. The impetus to treat this condition and the ingenious techniques devised to access the suppressed 'god-side' of man gave rise to religions.

These ancient traditions are mirrored by our scientific view of the past and present. Anthropologists tell us that our direct ancestors lived in the tropical rain forest – and our closest relatives, the fruit-eating apes, still do. Various disciplines, including climatology and palaeontology, have found that the evolution of many forms of life have been profoundly affected by repeated ecological catastrophes. And from the sciences of neurology and psychology we know that we have two distinct selves. The latest research in this field is now revealing that the dominant side is perceptually limited and continually makes up confabulated tales to cover its fractures of reality. The dormant side, in contrast, has exceptional latent abilities – even its capacity for pleasure is more encompassing.

It is a deeply held scientific assumption that humans not only represent the pinnacle of evolution but that advance proceeds apace. This view however is in conflict with observed behaviour – our inability to harmoniously coexist with each other and the increasingly rapid exploitation of the earth suggests we really are suffering from a psychological malady.

The two versions of human history – one poetic, contextual and right brained, the other analytical, rational and left brained – are remarkably similar. The only significant point of disagreement is degeneration verses advancement. Currently the advancement argument holds sway, yet evidence from both versions suggest that this interpretation is flawed. If the side that has reached this conclusion has been negatively modified in some way – can we trust its judgement?

CHAPTER ONE

~ψ~

Two Sides To Everything

An investigation into the characteristics of the left and right sides of the human brain reveals certain anomalies. Facts about how our brains work coupled with such oddities such as handedness, sleepwalking and religious experience are all clues to a 'second system' hidden within us that has a higher level of abilities than we realise. A new theory of how this may have arisen explains why accessing this second system is not straightforward. Over the course of our evolutionary history we have suffered a deleterious change in our consciousness that has affected not only how we act but also who we are. The consequences of this are far-reaching.

Research carried out over the last thirty or so years has quietly revolutionised our understanding of how the human brain works. Some of it challenges long held notions of who and what we think we are. We may be on the cusp of a new and very profound understanding of consciousness which, when it filters into mainstream culture, could transform not only ourselves as individuals but also how we organise our societies. This new perspective will, we predict, reach into the realms of enhanced health, supercharged immune systems and consciousness change. How can this be?

Science has come a long way but all too often conservatism and dogmatism can block the free flow of new ideas. Biology has been constrained by the fundamental beliefs that animals are survival machines run on chemicals, in particular DNA, and that the brain is a discreet organ that is also largely driven by chemistry. The brain is also regarded as the seat of consciousness and thus it follows that an individual human is a separate island of consciousness isolated from others and anything 'beyond'. Recent research has indicated that these foundations of physics and biology, though they have been in the past helpful to our understanding, are only partially true. In our everyday lives we may think of ourselves as being separate, however we all have the ability to open ourselves to a much more subtle level of experience – one of underlying unity and connectedness. These perceptions, often described in terms of religious experience, occur when something takes us out of our normal mode of brain processing.

These new insights are already finding their way into medicine. We are now discovering, for example, that the mind has direct connections to the immune system, raising possibilities of 'thinking ourselves well'. Renewed scientific interest in the phenomenon of autistic savants is also beginning to unlock the secrets of what were once considered to be supra normal abilities. All this new evidence has started to shift the orthodox scientific view. What is becoming clear is that the human mind has untapped powers. The new theories outlined in this book offer unique insights into these latent abilities, how they arose and why accessing them today is problematical.

THE DOUBLE BRAIN ~ψ~

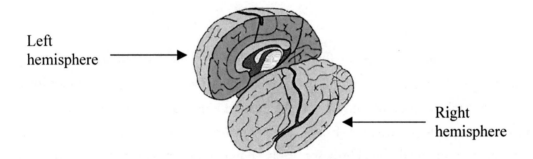

Left
hemisphere

Right
hemisphere

Fig 1a: the double brain

Before proceeding further we need to understand a few fundamentals about the structure and function of the brain. Animal brains are basically dual structures. As far as we know both sides are essentially alike in structure and function. There are differences but in humans and to a lesser extent some ape brains, these are more significant. Some change has occurred in our evolutionary history. Although the two hemispheres of the human brain look like reciprocal halves of a soft grey walnut there appear to be differences in what each can do. Certain functions, most notably speech, seem to be lateralised. Although our understanding is continually being upgraded, it is broadly accepted that logical concepts like time, sequence, speech and language are largely handled by the left side, and creativity, spatial awareness and pattern recognition by the right. Modern brain scanning technology has added further confirmation. People have been hooked up to scanners while performing certain tasks; reading and solving math problems lights up the left brain, whilst the right brain is more active when recognising faces and listening to music.

Despite feeling that we are a single unified self, our two contrasting hemispheres each have their own way of perceiving the outside world. The extent of these differences in perception is enough to suggest that we have two separate minds, and, in a way, two distinct selves within each one of us. The puzzle of why this should be, why the two sides of our brains function in such different ways, is something this book hopes to answer.

The early research on lateralisation was carried out in the 1860s by Paul Broca who noted that disturbance of language function in patients was linked to lesions situated in the left side of the brain. The discovery that a complex learned ability depended on structures on one side of the brain, led to the formulation of the concept of cerebral dominance; a term that means one hemisphere is predominantly responsible for a certain function. Further ideas, developed in the late nineteenth century, suggested that one hemisphere actually took precedence over the other.

The function for speech is located in the left hemisphere in 98% of right-handed people and over 65% of left-handers. As speech is closely associated with the way we think and reason, the left hemisphere became known as the dominant or major hemisphere. This view was reinforced when it was discovered that the movement of the right (usually dominant) hand was also controlled by the left side of the brain.

While the two hemispheres can co-operate with each other, each half contributing its particular abilities to the task at hand, there are times when the two sides are in conflict. If the two sides of a brain analyse any given situation in contrasting ways, one side may have to override the other to avoid internal interference. A democratic two-headed system makes for trouble or certainly confusion. For a variety of reasons we will return to later, the left has become the major player. It in effect keeps the right shackled. The fact that right hemisphere activities, such as singing, are inhibited in many people shows the extent of this interference.

Until relatively recently, the right hemisphere was thought to have only a minor role. Some human biologists even came to regard it as an almost vestigial organ, promoting the left as the storehouse of all complex and highly regarded abilities such as language, speech and logic. This idea was never biologically tenable however, for why would such a huge volume of metabolically expensive nervous tissue be maintained throughout the course of evolution if so much of it possessed no important function?

It is an interesting reflection on our culture that the right hemisphere has been for so long held in such low esteem, never reaching the high intellectual level conferred on the left by its possession of the capacity for language. Speech has always been regarded as one of the indicators of our high evolutionary status thus, the logic runs, the left hemisphere must be more evolved than the right.

In our western society we value speech particularly highly. We are excited by the first words our children speak and actively encourage them to name things. Then, for the rest of their childhood we tell them to 'hush the noise' and 'turn down the volume control'. Eastern, African and Native American peoples have a quite different cultural view of language. There is a much greater trust in reading a person's wants and intentions through facial expression and actions. There is a distinct preference to communicate with small children via body language and touch rather than words. Rather than relating from a distance (talking to the baby in a cot), these mothers will hold their babies close and feel their needs. Some peoples, such as the Gusii of Kenya, will talk to their babies much less than we do in the west. Their children will learn to speak at a later age than western kids do but they catch up in the end and the slow start gives them a chance to develop greater body awareness. Perhaps in our modern societies we have either lost or cannot easily access an ability to listen to the body due to the over dominance of the language system.

Free from the left

Advanced research techniques that are now coming on stream are beginning to reveal that the right hemisphere actually has the capacity for a much greater function than ever imagined. Ideas of left hemisphere superiority are falling by the wayside. The new orthodoxy is that both hemispheres have an equal level of complexity but differing functions. However there are clues that could tip the balance the other way.

It has been found that there are chemical and pharmacological asymmetries between the hemispheres. E.A. Serafetinides, working at the Guy-Maudsley Neurosurgical Unit in London, administered LSD-25 to patients who had undergone left or right temporal lobe removals. (It may seem extraordinary that humans can survive with only half a brain but there are medical conditions that necessitate the removal of part or even a complete hemisphere.) He found that the typical perceptual responses to LSD, psychedelic hallucinations and 'mind-expanded' states, disappeared after right but not left temporal lobotomy. This suggests that the drug did not have any affect in the left hemisphere. This is perplexing. Is the right side of the brain more sensitive than the left?

Even more startling research is beginning to show that the right hemisphere, if somehow freed or decoupled from the left hemisphere, has latent abilities that exceed what we usually regard as normal.

For example, one nine year-old boy was transformed from an 'ordinary' school child to a genius mechanic after a bullet destroyed a part of his left hemisphere. Ten year-old Orlando Serrell also acquired uncanny abilities after a baseball struck him on the left side of the head. After the injury healed, he found he could perform calendar calculations of baffling complexity and also recall the weather, where he was and what he was doing for every day since the accident. His feats made the news headlines.

In yet more cases, five patients from the Californian School of Medicine developed amazing drawing skills after dementia destroyed some specific parts of the left side of their brains. One of them had spent his life fitting car stereos and had never shown an interest in art. When dementia destroyed neurones in his left frontotemporal cortex, he suddenly started to produce sensational images depicting scenes from his early childhood. It was as if the destruction of those brain cells took the brakes off some innate ability that had been suppressed for most of his life.

These unlocked abilities parallel the astounding numerical, musical and artistic skills of autistic savants, memorably portrayed by Dustin Hoffman in the film 'Rain Man'.

There is a well-documented real life case in which this sort of heightened ability has reached a quite phenomenal level. Stephen Wiltshire is an autistic savant. He has severe learning difficulties yet, despite huge deficits, he has an extraordinary talent for drawing. At the age of eleven, he drew the Natural History Museum and other notable London landmarks to such a high standard that the well-known architect/artist Sir Hugh Casson described him as "the best child artist in Britain".

But it is Stephen's memory that is so astounding. At the age of 15, for a television documentary, he was taken for a half-hour helicopter ride over London. He took no notes or photographs and yet, back on the ground, he was able to produce a totally accurate aerial drawing of four square miles of the capital, incorporating over 200 buildings. His pencil never stopped, and he never corrected his work. It could be argued that London was already familiar to him; but he recently performed exactly the same feat in Rome, this time reproducing an accurate panorama of the entire city on a wall-mounted six metre long roll of paper.

Stephen Wiltshire's combination of photographic memory and highly accomplished draughtsmanship is an extraordinary, almost preternatural faculty. And yet it may be simply an extreme manifestation of a skill-set we all possess. It is possible that somewhere within all our brains there is the ability to mentally 'photograph' all the detail that we see. We normally filter out this detail, as it is either not relevant or too complicated for the dominant part of the brain to cope with.

Betty Edwards, in her popular book 'Drawing on the Right Side of the Brain', summarises the left-brain function very succinctly.

'The dominant left verbal hemisphere doesn't want too much information about the things it perceives – just enough to recognise and categorise. The left brain, in this sense, learns to take a quick look and says, "Right, that's a chair (or an umbrella, tree, dog, etc.)." Because the brain is overloaded most of the time with incoming information, it seems that one of its functions is to screen out a large proportion of incoming perceptions.'

Most experts assume that this left brain filtration system is necessary to enable us to focus on our thinking. The left brain just cannot cope with the mass of detail so, in order to function at all, it either has to filter most of it out or not process it in the first place. So where is a photographic memory like Stephen Wiltshire's stored? Almost certainly in the right hemisphere or at least facilitated by its operation.

One crucial finding about savants is that they often have structural and functional damage to their left hemispheres. This has led some researchers to suggest that these exceptional skills are released when the right hemisphere is able to work on its own, without any inhibition or interference by the left. One scientist who is looking at just this, Professor Allan Snyder from the Centre of the Mind in Australia, has found that 'switching off' the left hemisphere with a magnetic interference field can indeed improve right hemisphere skills such as drawing.

Stephen Wiltshire can look at a whole city and replicate it exactly; he can precisely draw a building with over a hundred windows, and yet he cannot numerically count, add or subtract. It appears he has lost the linear processing function of his left hemisphere, but that very loss has allowed his right hemisphere to flourish so spectacularly.

In everyday life, we all come across examples of the left hemisphere imposing a kind of rigid censorship on the right. Most of us know individuals who are otherwise intelligent and apparently high functioning, but have a total inability to sing, draw or paint. This complete lack of artistic ability is sometimes called a mental block, and it may literally be true. The block may be coming from the left hemisphere exerting too much dominance over the right.

Some have argued that left hemisphere dominance has been a necessary part of human evolution, enabling us to develop the full complexity of our language skills. But the latest research suggests this may be false. The right hemisphere turns out to have an equal capacity for language. For example, epileptics who have had parts of their left hemisphere surgically removed, don't cease to understand language; the right hemisphere takes over the language function. There is some evidence that suggests that the right hemisphere may be even better with language than the left. The autistic savant Daniel Tammet can not only perform extraordinary mathematical calculations at breakneck speeds but, unlike other savants who can perform similar feats, he speaks seven languages (French, German, Spanish, Lithuanian, Icelandic and Esperanto) and is even devising his own – "Mänti". Icelandic is a very difficult language to learn and yet Daniel Tammet mastered it within a week. Though there are many theories about savants, it is usual that some kind of brain damage causes the affliction – perhaps the onset of dementia later in life, a blow to the head or, in the case of Daniel, an epileptic fit when he was three.

Scans of the brains of autistic savants suggest that the right hemisphere might be compensating for damage in the left hemisphere. There is therefore the possibility that, in Daniel's case, his right hemisphere is giving him his outstanding language ability (as well as the ability to calculate cube roots quicker than an electronic calculator and recall Pi to 22,514 decimal places). If it is indeed right hemisphere processing that is giving Tammet his facility with language then, even if the right has compensated for a breakdown of the left, it somehow processes languages better than a normal left hemisphere can. Does this mean that the right has magically grown ultra proficient or are its skills inherent and normally kept bottled up by left hemisphere dominance? Time and research will answer this fundamental question.

Other unusual things happen to our language skills too when the dominant hemisphere is left without the support of the right. For instance, when the right hemisphere is irreparably damaged, speech is typically delivered in a monotone and even the difference between male and female voices becomes impossible to discern.

In one bizarre case, a nine year-old boy suddenly learnt how to speak after his left hemisphere was removed. 'Alex' was born with a disorder called Sturge-Weber syndrome, which disrupted the blood supply to the left side of his brain. He suffered epileptic fits and could only utter a few indistinct sounds; his only intelligible word was 'mama'. He was so ill that doctors decided that the only way his fits could be controlled was by removing the damaged half of the brain. Most remarkably, two years after the surgery Alex could talk like a normal child. This result caused quite a stir because, according to accepted theory, we can only acquire language during the first few years of our lives. Most people who start to speak unusually late never become very proficient. As we grow older it seems that our brains lose 'plasticity' – networks of nerve cells lose the ability to form new connections on which learning depends.

Most children with Sturge-Weber syndrome learn to speak in their early years even if the left side of the brain, the site of the so-called 'language centre', is damaged. The right side is able to take over this function. In Alex's case however, the fits or perhaps the drugs used to control the fits may have interfered with the right hemisphere. John Marshall, a neurologist at Oxford University, has proposed that Alex's right hemisphere may have actually acquired language ability long before the operation but the faulty left hemisphere masked the function. But alternatively, Mortimer Mishkin, a neuropsychologist at the National Institute of Mental Health, Washington DC and a member of Alex's medical team, argues that this case suggests that the brain may remain plastic for language until puberty. Both suggestions are intriguing. Firstly, it is clear that the right hemisphere can facilitate language and thus this is not necessarily a specialist function of the left. It is also probable that a damaged left hemisphere can interfere with the functioning of an undamaged right. Secondly, Miskin's suggestion raises the possibility that the plasticity of the left hemisphere, and hence its learning capabilities, is more susceptible to degeneration than the right and that there may be a differential response in this area to the body's steroid hormones that become more active at puberty.

This link with hormones may be pivotal. Testosterone is known to inhibit left hemisphere development. Most cases of autism are caused by some prenatal interference with brain development and it is thought this hormone may be involved, particularly as autism and savant skills are about six times more common in males than in females. It is possible that a slight shift in the balance of testosterone at a critical foetal stage may result in an over development of the right hemisphere and stunted growth of the left. It seems unlikely however that the same chemical is causing damage in one half of the brain and boosting the other. It is much more plausible that the damage to the left is reducing its ability to dominate the right, thus allowing functions of the right hemisphere free rein. All this raises many questions concerning the roles of hormones and hormone inhibitors on brain development that, as we will see later, are central to our argument.

For now, we may conclude that the left hemisphere is dominant, not because of any superior skills, but simply because it is suppressing the right. This strangulation may not only cause the right brain to become atrophied through under use – particularly in regard to its language skills - but also hide a range of other abilities.

While we have presented some initial evidence that raises the possibility that the left side of the brain may have limited function in comparison to the right, it is the left side that remains dominant. Over the course of our human development, something must have occurred to bring about this strange state of affairs that has rendered most of us right-handed but in the thrall of a piece of neural equipment with less than optimal function.

HANDEDNESS ~ψ~

Handedness is a unique feature of humans. It is a physical symptom of the dominance and preference of one side over the other. The body is mainly connected via the nervous system to the brain in a crossed-over way. Thus the dominant left hemisphere controls the right side of the body including the right hand while the right hemisphere controls the left side. Most people are right-handed because in most people their left hemisphere is the dominant one. However, this control can be overridden. In cases of damage to one hemisphere or even the removal of one side of the brain, the remaining half can take complete control of both sides of the body.

While most people are right-handed, there is a variation in the population. Some can use both hands to a greater or lesser degree, and some are genuinely left-handed. About 90% of people are right-handed and 10% left. The percentage of people preferring to write with their left hands has increased over the last century from about 2% in the 1930s to 11% in the 1980s. This change may be largely due to greater tolerance amongst parents and teachers, for in the past left-handedness was actively discouraged.

There are horrendous tales of attempts to correct this trait. Parents want 'normal' children. Former United States Vice-President Nelson Rockefeller was a left-hander who was encouraged to change. Around the family dinner table, the elder Mr Rockefeller would put a rubber band around his son's left wrist, tie a long string on it and jerk the string whenever Nelson started to eat with his left hand, the one he naturally favoured. As a result of this pressure, Nelson achieved a degree of awkward ambidexterity. He was fortunate to have as successful career as he did, because forcible handedness change can often cause serious problems such as a lifelong stutter, left/right directional confusion and dyslexia.

Although the use of a preferred hand is the most obvious difference between left and right-handers, it is not the only difference. Left-handers show less lateralisation than right handers. Thus left-handers more frequently process language and spatial information in both hemispheres than do right-handers. Mixing function in both hemispheres creates the potential for internal conflict, as the dual processors are more likely to interfere with each other. This may be the reason why left-handers are more likely to suffer from speech impediments and reading difficulties like dyslexia. Yet left-handers excel at certain tasks and professions, notably higher mathematics, chess, music, draughtsmanship and particularly art. Leonardo da Vinci, Raphael and Michelangelo are all thought to have been left-handed.

It appears from the degree of excellence in these fields that left-handers have greater access to the mode of processing carried out in the non-dominant hemisphere. Thanks to some very meticulous research, carried out in California by Roger Sperry and his students, we now know much more about how our hemispheres operate. Sperry has done groundbreaking work showing that we have two types of reasoning skills – verbal and non-verbal. As the term implies, verbal involves language, non-verbal involves other abilities such as pattern recognition and spatial awareness. He has demonstrated that verbal and non-verbal skills are controlled by the left and right hemispheres respectively.

One of Roger Sperry's students, Jerre Levy, went on to show that the right brain's mode of processing is rapid, whole-patterned, spatial and perceptual – very different from the left brain's linear, verbal and analytic approach. But each, he showed, are equally complex. Thanks to such work, our view of the brain and its capacities has changed markedly over the last thirty years. The right hemisphere is now known to possess highly complex functions and it certainly is no longer regarded as vestigial.

As an aside, Sperry points out that our education system, as well as science in general, tends to neglect the non-verbal form of intellect. Exams, for example, predominantly test verbal skills, neglecting other types of non-verbal intelligence. This helps to explain the many cases of career high achievers having a much less distinctive school record.

The degree of preference for one hand increases throughout childhood. Young children are generally ambidextrous but from around the age of four one side becomes more dominant. Lateralisation is usually complete by the age of ten and this is marked by changes in how the brain operates. From an analysis of children's art we know that, at around ten, the concept of what a thing looks like takes precedence over direct observation. The left hemisphere's way of processing based on classification, words and symbols takes precedence over spatial and holistic perception. It appears that the activity of steroid hormones, which increases as we get older, plays a part in these changes.

Two of the leading researchers into cerebral dominance, Geschwind and Behan, have found a link between the learning difficulties of strongly left-handed individuals and excessive production or sensitivity to testosterone in the foetus. Research in this field appears to show that an imbalance of this steroid can cause the diminution of the rate of growth of the left hemisphere, as well as suppression of the development of the thymus gland. These two factors could account for the higher frequency of left-handedness and learning disabilities in males. This is startling enough, but the discovery that testosterone and other steroids are playing a significant role in the development of behavioural asymmetry may be profoundly significant. It is quite possible that changing our steroid regime could not only change our physiological function but even the way we think.

Accessing the right brain

The capacity to access the right hemisphere mode is not limited to a few gifted artists. We have all been there. Whilst day dreaming, we are in at least partial right hemisphere mode. We access it when we are absorbed in activities like playing instruments and listening to music, when we are lost in painting and unaware of the passage of time and when we are in a beautiful landscape and it seems that time stands still. Moments like these have been termed 'peak experiences'.

These special states are characterised by one thing – the absence of the seemingly ever-present chatter from the left brain. It is only when that constant verbal dialogue with oneself ceases that peak experiences are possible. That is why activities such as painting, music and even fishing are so restoring and why meditation can be so effective.

Betty Edwards has noted that quietening what she calls the 'L-mode' may partially explain how practices, such as meditation and shifting consciousness by fasting, chanting and taking drugs, work. She also comments that drawing itself can induce a changed state of consciousness that can last hours, bringing significant satisfaction.

Why does reducing the influence of our left hemisphere in favour of our right have such a positive effect? If the left hemisphere is so much the complete and sophisticated side of our brain, why is it apparently such a relief to escape its mode of function and flee to the right side? There is something of fundamental importance here.

Is handedness helpful?

An equally simple but often overlooked question is why do we not have two fully functioning hands? There may be certain advantages, or certainly no disadvantages, for having a preference for repeatedly using one hand for certain tasks but would it not be better for us to be more ambidextrous? What events or pressures in our past could have given rise to this strange human anomaly? Is it likely that spear throwing (or something similar) was enough to cause the establishment of this trait? Did women throw enough spears too?

Such fanciful ideas have been bandied about to explain handedness but the fact is there are no plausible theories to explain handedness in terms of evolutionary selection pressures. It is more logical therefore to assume that handedness is a by-product, and not a very useful one, of cerebral dominance. Because we blindly assume that humans are now at their highest development, handedness is regarded as a positive trait – something inherently human and so, naturally, good. But it could be argued that handedness is a symptom of dysfunction – an indication of a flaw in the human make up.

Interestingly, there is some evidence that deliberate manipulation of normal brain function can shift the body towards a greater ambidexterity. For example, significant alterations of handedness occur when the brain is deprived of sleep. Sleep deprivation experiments have shown that the left hand (in normally right-handed individuals) can be used with increasing ease and ability with increasing hours of sleeplessness. There appears to be a cross over point in which the mind is confused as to which hand is dominant but this then settles into a more balanced ambidexterity. It becomes easy to move the hands and arms in synchronicity. The lack of sleep then seems to suppress the left hemisphere allowing an even-handed state to emerge. This disappears as soon as normal sleep patterns are resumed, re-establishing the dominance of the left hemisphere.

The celebrated French ear nose and throat specialist, Dr Alfred Tomatis, noted a similar effect whilst conducting hearing tests among workers at an industrial factory. The factory employees were engaged in precision work on jet engines that led to high levels of fatigue. The majority of his subjects were right-handed. Tests showed that, when fresh at the beginning of the week, they all displayed their natural right-handedness. But by Thursday and Friday evening, before they quit work, they showed mixed laterality with neither hand dominant. They also became hesitant to talk to anybody – another indication that left hemisphere dominance was not so complete as usual.

There are further clues that link handedness to left hemisphere dominance. Under hypnosis a more ambidextrous ability can emerge, and something similar is seen too in individuals with multi-personality disorders. It can take just a second or two for those with multiple personalities to switch. In the same instant they can switch handedness too. Right-handed sub-personalities turn into left and vice versa. As any right-handed person who has tried using scissors with their left hand knows, handedness is not something one can normally flip on and off. It has been regarded as a genetically fixed trait but this cannot be correct if it is affected by factors such as hypnosis and tiredness. If handedness is not a result of human evolution then perhaps it is a symptom of some sort of imbalance or malfunction.

SLEEP ~ψ~

Scientific literature suggests that the lack of sleep can be dangerous. Medical evidence indicates it can cause psychosis and even death but this is by no means the whole story. Research paints a less than clear picture. It was reported in the Daily Mirror (24.02.98) that a 55 year-old Vietnamese woman had not slept for 38 years. Her sleeplessness was initiated by trauma but she now claims to be well, doing gym exercises every morning and never feeling tired. Her doctors are baffled because everything they tried, including heavy doses of sleeping drugs, failed to make her sleep.

Dr Tomatis believes that the need for sleep is exaggerated. His findings suggest that the cortex needs constant energy inputs via sensory intake and, as most people don't have enough mind stimulating activities, they turn to sleep as an escape and a refuge. Similarly, James M. Kruger, whilst assistant professor at Chicago Medical School, stated: "For all we know, we don't need sleep. If we had a drug that blocked the effect of the sleep factor in the brain, we might be able to stay awake twenty-four hours a day without ill effects."

Reducing sleep has been used as a positive tool in many spiritual contexts to alter states of consciousness. Many religious traditions advocate the use of short or extended periods without sleep in order to access spiritual insight and such practice is even hinted at in one of the most ancient texts of all – the Sumerian Epic of Gilgamesh:

> "*Who will assemble the gods unto thee, that thou mayest find the life which thou seekest? Come, do not sleep for six days and seven nights.*"

The hero Gilgamesh, in his quest for immortality, is challenged to defeat sleep, the younger brother of Death, by not sleeping for six days and seven nights. We are told that he failed – as soon as he squatted down on his haunches, sleep, like a fog, breathed over him. But the very fact that conquering sleep is mentioned at all in the context of attaining a spiritual goal is significant, particularly as this is one of the earliest stories ever committed to writing. This tradition is parallelled in the Vedas. In these primary Indian scriptures we are told of all night rituals in which participants drank 'soma' – an invigorating drink that killed the demons of the night, prevented sleep and brought about sun-like visions of the gods. Perhaps the ancients knew something we are only beginning to rediscover.

Sleep deprivation and spiritual practice come together in some form in most religions. Buddhists regularly engage in all night periods of meditation (linked to the lunar cycle) and most monastic traditions have some restrictions on sleep, usually starting observances before dawn. The Buddha himself attained enlightenment after, according to one account, spending seven days and nights awake, in deep meditation, under the Bodhi tree. In North America too, the vision quests and sun dances of the indigenous people entail days and nights of continual practise. Despite the general perception that plenty of regular sleep is necessary for normal function, such a range of clues suggests that this may not be strictly true. From sleep walking (a state in which some part of the brain is asleep whilst another part is active and functional) and the deep insights that can emerge through dreams to the extraordinary outpourings of the sleeping prophet, Edgar Cayce, many anomalies remain unexplained. Could it be that an explanation may lie in a differential requirement for sleep between each half of the brain?

If the right hemisphere needs less sleep than the left then it follows that a normal level of sleep is necessary to maintain the left hemisphere's function, including its dominance. Reducing sleep may therefore be a means of exploiting this weakness. By starving the left hemisphere of its recharge time for long enough to run its 'batteries' down, its overall function may decrease along with its ability to suppress and maintain control. The opportunity to stimulate and reengage the potential of the right hemisphere, whilst free from this control, may be the basis of many powerful techniques used, in the past and present, to achieve higher consciousness function. If perceptual experiences such as religious oneness, bliss and so-called 'spiritual' abilities such as clairvoyance lie locked somewhere within the capacity of the right hemisphere then accessing this by using sleep deprivation to reduce suppression makes eminent sense.

One of the authors (A.W.), for over a decade, has personally experimented with up to eleven days and nights of sleep deprivation. On one trial (over eighty hours spent awake) he experienced an extraordinary change of perception that he describes as 'all-encompassing religious bliss'. He believes that for a period of about twenty minutes the left side of his brain went to sleep leaving the right side awake and functioning freely. This conclusion was reached from other incidental affects such as an initial block that prevented him from talking. After some internal mental investigation, that seemed far from rational or linear, a few grunts and broken syllables started to emerge. And when the speech mechanism was restored, there were several interesting differences. The voice sounded more resonant than normal. There was expression without the precursor of thought and there was some poetic element to its structure too.

When a more normal sense of consciousness returned, the quality of the voice changed and so did the syntax. The euphoria dampened a little too. There was still tremendous joy but, in parallel, there was a sense of confusion and bewilderment. Something wanted to know what had been going on. The self that had been to sleep felt it had missed out on something. When it woke up it wanted to find out who had been running the show.

This experience had knock-on effects. Some changes in perception were apparent. There was a temporary ability to simultaneously follow two conversations without switching between the two. Pliny writes that Julius Caesar could do something similar – he could dictate up to seven letters at one time and simultaneously listen to four or five reports from different regions of his expansive empire. Also, on one day, and for one day only, from the moment A.W. awoke, all thoughts that occurred were in rhyme; a most peculiar phenomenon. The process itself was completely automatic. Even when consciously trying not to think in rhyme, these thoughts too stubbornly emerged in rhyme. And the structure and phrasing of the thoughts again had a poetic quality. It was extremely interesting, but unfortunately by the next day it had gone. The muse had left.

There is historical evidence that rhyming ability is a right hemisphere function. Plato called it 'divine madness'.

'All good poets, epic as well as lyric, compose their beautiful poems, not by art, but because they are inspired and possessed [...] there is no invention in him until he has been inspired and is out of his senses and the mind is no longer in him.' (Plato, Io)

In early Greece, the epic poets and muses appeared to have the ability to tap into 'unwearying flows' of songs. The words came directly from 'the source' without artifice or rational invention. The normal thought process, the slow linear reasoning of the left hemisphere, is bypassed in this mode. There are hints from the earliest of writings that accessing this mode of function may have been much more common within past cultures. Perhaps the dominance of the left hemisphere was not quite so rigid in those days and this allowed access to a different set of functions and abilities.

It appears that by limiting the amount of our sleep we may, under certain circumstances, be able to reengage these lost abilities.

Hidden Potential

Over the following weeks, as A.W.'s sleep deprivation experiments continued, other abilities emerged, which were consistent with greater access to right brain functioning. The first was an ability to juggle, coupled with a change in visual perception that allowed the track of all the balls to be followed simultaneously. Not a huge scientific breakthrough perhaps, but interesting nevertheless because juggling does require a complex level of co-ordination. It is the sort of task the left brain with its step-by-step mode of functioning has great difficulty with. This type of ambidextrous ability was tested at a later date during more formal sleep deprivation trials.

Another associated change noted was an enhancement of peripheral vision. Number plates of cars could be read as they sped by and full on vision was enhanced too. When looking at trees, the perception incorporated all the leaves and the shapes of the leaves in detail and at once. There was a heightened awareness and clarity. It seemed that the filter the left brain imposes on perception was beginning to be by-passed.

Around this time, childhood memories were experienced at a profound and exquisite level too. Details of the patterns of carpets and curtains, smells and even emotions felt at the time were almost fully brought back to life. Accessing such vivid memories, though unusual, can occur spontaneously. Some shamanic traditions hold that we all can remember everything that has happened to us in great detail. They tell us that if certain techniques are engaged we can virtually relive events from our past. These techniques, that in some respects parallel meditation, quieten the left hemisphere's sense of self and by sweeping the head from side to side engage the right to allow the deeply buried memories to emerge. This procedure indicates that the most functional and deepest part of our memory lies either within our right brain or is accessed via its function. Because our (less efficient) left brain is dominant, we routinely experience difficulty accessing more than our superficial memory. We are dominated by a part of our brain that either filters out the detail or doesn't retain it.

Russian researcher, V.L. Raikov, has noted similar detailed memories of early childhood during experiments using EEG monitors on subjects undergoing hypnotic age regression. He observed striking activation of infantile memory, including characteristic brainwave patterns during the hypnotic regression. Brian Lancaster in his book 'Mind, Brain and Human Potential' argues that these memory feats are due to the attenuation of the subject's sense of 'I' that is removed by the hypnosis. 'In other words, our infantile memories may be extensive but we cannot access them because they were laid down before the 'I' formed. As adults we can only recall that to which our current 'I' has connection.' If the 'I' is a product of the left hemisphere function and becomes established with increasing left hemisphere dominance in early childhood, this begins to make sense, particularly when we realise that hypnosis, like dreaming, accesses our right hemisphere mode of function.

Could it be then that, despite its achievements, our normal dominant self somehow limits optimum function? Take the subject of reading for example; why can we not sweep the page and take in all the information at once? Our usual reading experience suggests that one word at a time is all that we (or at least our left brains) can manage. Yet there are courses one can take to learn how to 'photoread' and with this acquired skill pages can be scanned in seconds and the content recalled. It is possible that attempting to take the information in by rapid visual scanning overloads the left hemisphere. It is being asked to perform something that it cannot do. Not being able to cope, it gives up allowing the right hemisphere to take over. Within the right brain mode of function much more detail can be absorbed within an instant but the trick is then to be able to access this information at a later date. The right hemisphere perhaps works a little like a computer scanner that loads information into the memory as visual images. As the left cannot translate all the visual information, what is then needed is software to decipher the 'photographs' of the text into words. It is all in there but new pathways need to be constructed to gain access to it. 'Photoreading' courses use various activation techniques to enable connections to be made between what they regard as the inner mind and the conscious mind. Mind-mapping, engaging dreaming, acting, discussing and drawing are all used to bring the content through into our accessible brain system.

On occasions, we can naturally click into something very similar to the 'photoreading' ability. A.W. has found it possible to read two lines at a time, and by pushing this, has on occasions increased his scanning level to blocks of six lines and even whole pages. After processing a page in this way he 'knew' what was written there but to bring it out rationally took a bit of time. It needed something like translation to enable it to come out of the rational side. Kim Peeke, a celebrated autistic savant, has an even better reading facility. He can read two pages of script in about eight seconds and can recall all of it perfectly.

A similar heightened ability can be accessed with sound. Occasionally a piece of music can be heard with 'different ears' and every note and phrase takes on heightened clarity, relationship and meaning. Perhaps all it takes to tap into all these latent functions is a suppression of left hemisphere dominance. If we can develop and expand the techniques for doing so maybe we can all gain glimpses of a different kind of mental processing. Reading pages of text in an instant and listening to several conversations at once may be just the tip of a fascinating iceberg. Right hemisphere function may include telepathy, direct knowing, a highly developed intuitive ability, enhanced immune function and, most importantly, a wholly different sense of self.

Telepathy

Of all the many extraordinary abilities and powers that humans apparently possess, one of the most controversial yet fascinating is telepathy. Many scientists dismiss the phenomenon but subtle two-way communications have been well documented between mothers and children, twins, sisters and others near relatives. This ability has been kept alive to a much greater extent in the Australian Aborigine population and by the African Bushman. Aboriginal people prefer to walk together in total silence because whilst walking they speak in the ancient way with 'telepathy' rather than voice. When they feel a sense of anxiety, they may sit down and enter a meditative state. Each part of their body is associated with a relative or a geographic location. When a part twitches, by touching that place they can visualise what is happening to the relative or at the location. These practices have been repeatedly observed by anthropologists to be accurate and effective. Perhaps the main reason this means of communication is not experienced more widely within our western culture is because it is blocked by the business in which we are usually immersed. We may be convinced too that these functions are beyond the normal range of human ability – thus any unusual experiences tend to be dismissed. However in such cultures where these communications are commonplace, access just takes concentration and the ability to still the internal dialogue. This is something we very rarely do but by reducing the internal voice left hemisphere activity will calm down thus allowing more function from the right hemisphere to emerge.

There is some scientific research work, carried out in Mexico by J. Grinberg-Zylberbaum and J. Ramos, which lends support to these experiences. Pairs of volunteers in separate rooms were asked to tune in to each other's presence. When they focused, not only did the brain waves (measured by EEG readings) synchronise but the electrical activity within each hemisphere of the brains of each participant also synchronised. It was further found that the participant with the most ordered or cohesive brain pattern tended to have the most marked influence on the other.

Other research carried out on staring provides even more evidence that we can have a physiological effect on one another from a distance. Rupert Sheldrake, in his book 'The Sense of Being Stared At', has amassed a database of more than 4,500 case histories of apparently unexplained perceptiveness that provides convincing grounds for the veracity of a wide range of psychic phenomena. He includes such senses as 'telephone telepathy' (when we think of somebody just before they phone.) We have all done this at some time, as well as experiencing the slightly uncomfortable awareness of having someone surreptitiously staring at us. Ingenious experiments carried out by Braud and Schlitz investigated this latter phenomenon. The results suggested that staring and non-staring periods could indeed be distinguished by physiological responses in the person being stared at. Braud and Schlitz also found that for the most part the detection didn't reach conscious awareness.

It would be intriguing to repeat some of these experiments with the additional parameter of using sleep deprived subjects. Would their levels of sensitivity be increased with less left hemisphere interference?

THE MANCHESTER TRIAL ~ψ~

To investigate some of the issues around sleep and enhanced body/brain function, a pilot study under Professor David Collins took place in September 1998 at Manchester Metropolitan University. Two subjects (including A.W.) stayed awake for five days and four nights while being tested and monitored round the clock. One further element to the experiment was that the two individuals had for a number of years been maintaining an almost exclusively raw diet, rich in fruit. We will be returning to the chemical and evolutionary significance of ancestral diets in the next chapter, but this unusual combination of factors meant that this experiment was unique.

It was expected that the longer these two subjects were deprived of sleep the more they would exhibit decreases in co-ordination and functional ability, however this did not occur. In fact some abilities actually increased as the experiment progressed. Professor Collins, who was trained in the military to deal with the affects of sleep deprivation, was surprised for this was not what he had found before.

During the five-day experiment stamina, physical abilities, co-ordination and mental responses were investigated and breath (gas analysis), heart rates and brain activity (EEG) were monitored. These trials were repeated at three hourly intervals around a 24-hour cycle and, once a day, brain activity was further checked using co-ordination trials. These tested the response to written instructions that appeared on a screen. For example the word 'green' would flash and the response would be to hit the green button.

Physical tests included jumps to measure height reached, and bouncing a ball against a wall, catching it alternately with each hand to measure co-ordination (this was timed). There were specific tests for left and right hands too. One involved putting pegs in a board; the task was timed for each hand. The results for all these tests were, from a standard viewpoint, unexpected. For instance, in the pegboard experiment initially the right hand was quicker but as the experiment proceeded the left hand improved its performance so that overall it actually achieved the faster times. Another test involved balancing on a 'seesaw' device. This was difficult and to begin with the plank just crashed from one side to the other. It took much effort to achieve any sort of balance at all but on the last day of the trial A.W. stood up and balanced perfectly. This puzzled Professor Collins for he had found in the past that balance was one of the abilities that markedly decreased with tiredness.

Unfortunately, to date, the EEG data has not been analysed. It would be extremely interesting to see whether the EEG picked up any progressive differences in left and right hemisphere activity over the experimental period. However, the overall results from the other tests show dexterity, strength and co-ordination increased rather than declined. It appeared that sleep deprivation in conjunction with a raw food diet was responsible for an unexpected and anomalous result.

Previously Professor Collins had worked with athletes who had achieved highly unusual and enhanced 'once in a lifetime' performances. These athletes described their mental state at these times in almost mystical language. They reported transcendent, fluid and flowing states. As such descriptions are more often associated with 'religious' experiences that occur when the left hemisphere's influence is reduced, it might indicate that the phenomenal performances could have involved a similar neural crossover.

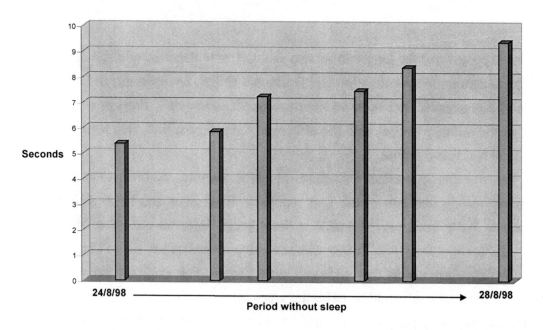

Fig 1b: This test involved the subject standing on a seesaw device and trying to keep as near level as possible. The graph shows the average time per test the subject stayed within a specified range of balance. It can be seen that ability increased as the trial progressed.

Time(s) to complete dexterity test - A.W.

□ Pegboard Right Hand ■ Pegboard Left Hand

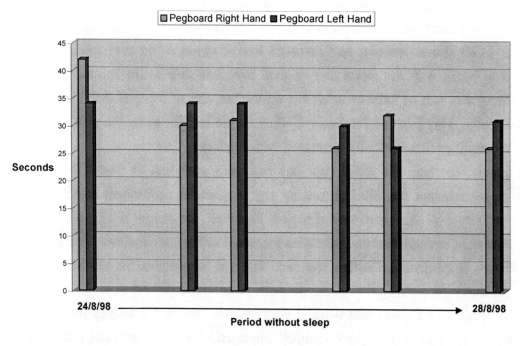

Fig 1c: This graph shows the time taken by each hand to remove, turn upside down and return small wooden pegs to their respective holes. This trial tested dexterity, hand and eye co-ordination and concentration. It can be seen that the time taken to complete the task decreased with the numbers of days spent awake. It may be worth noting that this normally right-handed subject showed an increasing degree of ambidexterity or even left hand superiority as the trail progressed. This accords with the observations by Dr A. Tomatis of the factory workers who became more ambidextrous as their fatigue levels increased.

The sleep deprivation experiment provides additional evidence for this thesis – higher performance is accessed when the right brain takes over from the left. It appears that the nutritional input was a factor too. Could the biochemistry of something that approached a human ancestral diet have helped reactivate right hemisphere function?

In another sleep deprivation study, Sean Drummond of the Department of Psychiatry, University of California, found further evidence that indicates the right hemisphere, under certain circumstances, is more resilient than the left. This experiment was set up to assess the impact of sleep deprivation on neural activation during verbal learning. Even though the duration of the trial was limited to 35 hours and the biochemical factors of the Manchester study were absent, the results still showed that reducing sleep changed the area of brain activation – there was a reduction in left hemisphere activity and a corresponding increase in the right. This was particularly surprising as the test was verbal. It seems therefore that the so-called verbally specialised left hemisphere may not be so specialised when suffering from a lack of sleep.

A NEW HYPOTHESIS ~ψ~

From research areas like these, an extraordinary hypothesis has emerged which is finding connections in many different fields and disciplines. The fact that it appears to fit into such a wide range of subject areas, and indeed links and clarifies them, adds credence to the basic theory.

Let us assume that sometime in the past the neo-cortex was effectively a single consciousness system: that is, there were no marked structural and functional differences between the two hemispheres. It was just one whole brain. At some point in time, however, something went wrong and damage was sustained to this highly sensitive system. This resulted in a progressive change in the most delicate structural components of the brain and this, in turn, changed the very nature of man's experience – it altered his consciousness.

In simple terms, the evidence we will present suggests that the human brain has suffered a significant long-term decline in structure and function. The damage is primarily restricted to the dominant half of the brain. This has created a distorted experience not so much of the outside world, but of the inner world – our very feelings of who or what we are. This is a mental state that is extremely hard to escape from. When the escape does occur it is usually brief and transient, but it can be profound, often being described in terms of bliss or religious ecstasy. A very different sense of self can emerge in conjunction with these experiences – a much greater, all encompassing sense of oneness that is described in some

21

form by all religious traditions. Many of us may have some inkling of this other state but it is unfamiliar and far from our 'normal' consciousness. That it exists at all, and is almost always regarded as positive, could indicate that the equipment we use to assess how we feel or what we experience is flawed. As our immune system, hormone system, and just about everything else is effectively run by the brain, this represents a somewhat disturbing picture.

There is some evidence that suggests that one of the primary causes of this damage is an altered level of testosterone activity. Testosterone is one of the most powerful growth-regulating hormones and is produced by both males and females. We will look at the ways (what we regard as) abnormally high levels of this hormone have worked this havoc in later chapters. For now we only need to note that the damage starts and is at its greatest in the earliest phase of development in the uterus, but it continues throughout our lives, rapidly accelerating with puberty in males and, to a lesser extent, menopause in females. Damage may be slightly greater in males as more testosterone is produced. It is also pertinent to note at this stage that the activity of steroids like testosterone are ameliorated by biochemical elements that were present within our ancestral diets and are still present in fruit.

If there is the slightest possibility that our self-perception equipment is indeed damaged, then not regarding its repair as an absolute priority would perhaps be the most telling symptom of the severity of that damage. It is quite possible that with a restored consciousness system, perception would be more vivid, tastes enhanced, vision brighter, sounds clearer and we would become far more joyful. It would feel sensual just to be in the body and our physiology, from the immune system to digestion, would run more efficiently too. For mothers, even giving birth could be blissful and pleasurable. We might find we had other abilities too. In a study on the mystic traditions of the aborigine people, the eminent author and anthropologist A.P. Elkin gathered compelling examples of higher 'psychic' powers. These included clairvoyance, telepathic communication, remote viewing, psychic healing, and journeys to other worlds. The fact that all these have been reported from other cultures too indicates they are part of the human repertoire. Instead of regarding such powers as abnormal perhaps we should ask whether our inability to access these things today points to a genuine decline in our functioning.

Is it possible that we really do possess an alternate self that has a greater range of abilities? In the following chapters we will be investigating the evidence for such apparently outlandish claims, and exploring how in the course of human evolution we came to gain and then lose these abilities.

CHAPTER TWO

~ψ~

From The Forest

The 'Left in the Dark' theory offers persuasive new explanations for the rapid enlargement of the human brain, our hairlessness, the length of our childhood, why we walk on two legs and our aggressive nature – questions that have perplexed researchers since Darwin. Strong evidence is also presented that shows for a crucial period of our evolutionary history our distant human ancestors were primarily forest-dwelling fruit-eaters, not animal hunters, as is commonly supposed. Archaeological research, primate nutrition and human anatomy all point to the same startling conclusion.

Primates evolved, diversified, came, went, lived, still live, reproduced and ate in the forest. Primates have arboreal origins and, in the very distant past, the lineage that gave rise eventually to the higher apes switched their diet from an insect-based one to one based on flowers, leaves, shoots, nuts and fruit. This may have happened in the region of 70 million years ago. Primates in general have certain traits that include a larger brain to body ratio than most other mammals. Can this be linked to this change of diet?

Other groups of animals show a similar correlation. Fruit bats have a larger brain/body ratio than their insect eating relatives (up to twice the brain size) and so do parrots. The intelligence of these birds has led researchers to jokingly classify them as 'honorary primates' because their ability to categorise objects and grasp abstract concepts like the similarity of shape and colour rivals that of chimpanzees. Species of primates that have a high percentage of fruit in their diet tend to have proportionately larger brains than do their cousins that eat a more leafy or omnivorous diet. These examples clearly show that changing from an insect-based diet to a fruit one is linked to an increase in brain size.

The primates that ate more fruit and came to depend on fruit would of course have to live in the forest because this is where fruit grows. In the warm wet climate of the tropics there is little seasonal variation and so trees such as figs could (and still do) provide fruit all the year round. Along with nectar, fruit by its very nature is a 'free lunch' provided by the plant kingdom. It is designed to be eaten, its quality as a food is a reward for seed dissemination and it can be gathered with relatively little effort. It is the most obvious food source for primates and hominids, and, as we will detail later, the fossil evidence of hominid dentition strongly suggests fruit was the most important element in their diet. Fruit is a very good food source. It is rich in nutritive value, easily digested and low in toxicity. The normal mechanisms of evolution would have acted on both primates and tree species so that they became adapted to a life of inter-dependence.

It is of benefit to the fruit trees to make the fruit as healthy a food as possible. Healthy primates mean better seed dissemination – thus there may have been degree of co-evolution at work here. If certain fruit, for instance a variety of fig, not only tasted better but made the hominid feel better then this variety would have been selected more often and hence dispersed more efficiently. (This scenario is entirely plausible for figs contain chemicals that elevate neurotransmitter activity, which can in turn affect a hominid's mental state.) If the chemicals in the fruit also enhanced neural expansion, a feedback mechanism may have been initiated; more fruit means more fruit biochemistry means more neural expansion and more fruit dispersal. Selecting only the more powerfully loaded fruits would fuel this process even more. The hominids may have unwittingly managed the forest environment by selecting and dispersing the most beneficial fruits. With more chemically rich fruit available, the more fuel there would be for acting on the hominid's neural system. A mechanism such as this could have led to a very rapid evolutionary leap.

There would have been dozens of different lineages and dozens of branches on the primate tree of evolution; some are known, some are waiting to be discovered and probably many more will be forever lost in the mists of time. But the key ingredient on the branch or group of branches that lead to the hominid line was a dietary specialisation on fruit and this necessitated a life in the forest. Unfortunately for the hominid researcher, animals do not tend to fossilise in tropical forest environments. This is a point worth emphasising. If no fossil hominids have been found in tropical forest environments it does not mean they were not there. No fossil chimps and bonobos have been found in this environment either and they definitely were there, and still are, eating a mainly fruit/plant based diet.

The current model of human evolution is based on a very limited number of fossil finds, while the fossil record of the great ape ancestry is almost entirely absent. (In fact there is not enough fossil evidence from human and hominid lines to categorically prove this *or any other* hypothesis of human evolution.) This gives credence to the hypothesis that the fossil remains of hominids that have been discovered represent populations that left the forest, but it in no way discounts the notion that a source population remained in the forest. If no remains of the forest apes survived as fossils, why should remains of forest humans? The fossil remains that we do have are likely to represent humans that had spent many generations out of their optimum environment and as such cannot give a true representation of what was really happening in the forest. The rather simplistic theory of hominid descent, that is implied rather than stated, is that the Ramapithecines were the ancestors of the Australopithecines that were the ancestors of the Homo line. Within the Homo line there was *Homo habilis* that was the ancestor to *Homo erectus* that in turn was the ancestor of *Homo sapiens*; 'A' leads to 'B' leads to 'C' etc. We suggest another pattern that maintains an evolving population within the prime forest area from which offshoots emerge. Species 'A' in the forest evolves into species 'B' which remains in the forest but an offshoot of 'A' emerges onto the savannah and its remains are fossilised and later given the name *Ramapithicus*. An offshoot of species 'B' leaves the forest and its fossilised remains are given the name *Australopithicus*. Thus the few fossil remains of hominids that have been discovered represent waves of evolutionary 'dead ends' (as far as continuing brain expansion is concerned) that left the forest. Many different lines of temporarily successful hominids could have followed this evolutionary route. Meanwhile the hominid lineage that lead to *Homo sapiens* continued to develop back in the fruit-rich, leafy and womb-like environment.

We propose that from the earliest days of primate fruit eating, the chemicals included in the diet started to short circuit the normal evolutionary mechanism. The chemical cocktail that is present in fruit started to modify the way our genetic blueprint was actually read and interpreted. (This process is known as *transcription*.) Powerful biofeedback loops were created which affected assimilation and biochemical function. DNA transcription would have been affected directly, very gradually at first, but at some later point a critical mass was reached which initiated rapid and profound changes. Juvenility was extended and brain size increased markedly as a result of this new biochemistry.

This was not merely a nutritional effect that fluctuated with season and whim. A specialist fruit diet provided a small but constant flow of these active bio-chemicals. They were present all the time, through countless generations, and significantly the chemicals would have influenced foetal development during pregnancy. As far as the primate's internal chemistry was concerned, it could have been an internal gland producing these physiologically changing chemicals for they were so continually present.

These primates were plugged into the tree biochemistry continuously and over millions of years. (External oestrogen-like chemicals in our present environment are causing biochemical changes in both humans and other animals today. This is a similar effect and confirms this mechanism is not merely theoretical. For example, some river fish have been found to become hermaphrodite in response to pesticide run-off, and it is possible that the sperm count in men is dropping due to residues of birth control pills that end up in our water supplies.)

Each genetic line would respond to this continual biochemical intake in a different way. Each line would be unique – a subtle variation on the model. There could have been hundreds of genetic 'starting points' and hundreds of different outcomes. Specialising in fruit eating is not a magical route that is going to automatically result in the development of a brainy super-hominid. But as traits such as increased brain size and extended juvenility are seen to a greater or lesser extent in the higher apes as well as hominids, it appears that the correlation between brain size and fruit eating is strong.

JUVENILITY AND BIPEDALISM ~ψ~

The continual assimilation of the chemicals found in tropical fruits would have had a direct affect on the mechanisms of juvenility. If there was an inhibition of the internal mammalian hormones that normally built up to a level that triggered sexual maturation, profound and rapid change could have occurred. If, for instance, sexual maturity were delayed for a year or two a very different animal would emerge for the whole neuroendocrine system would have had a longer period to grow and develop beyond its previous parameter. This hominid with a longer juvenile period would be different to its ancestor that wasn't eating this chemically rich diet. Though the neuroendocrine system (the system ultimately controlled by the brain that produces hormones) itself plays a central role in the regulation of juvenility, the forest fruit, we hypothesise, introduced a gremlin into the works. It effectively caused an extension of the juvenile period as well as changing the very mechanism that regulated the juvenile period. This led to a fast track of evolutionary change.

Could this have actually happened? The evidence suggests that it may have done. We know that within humans today, the period of rapid brain development ends at puberty. It would have been no different in our distant past. We also know that the chemicals present in fruit have the ability to suppress the steroids that induce sexual maturation. Thus their action could have created a window of opportunity for additional brain development. This would have resulted in a primate with a modified neural-endocrine system, for a bigger or more developed brain would run this system in a slightly different way. The progeny from this hominid with the bigger brain would in turn be exposed to a changed internal chemical regime. Not only would the foetus be exposed to a changed environment in the uterus (because its mother has a modified neural-endocrine system), but also the resulting brain would pump a

26

slightly different complex of chemical messengers itself. The effect could therefore build with every succeeding generation creating a lineage with an increasingly modified biochemistry, which would build individuals with an increasingly different structure and function, even though there had been no change in the DNA codes.

So the diet of forest hominids could indeed have caused an extension of the juvenile period and this would have allowed a longer period in which the brain could grow and develop which would presumably have provided scope for enhancement of function. The pressures and opportunities that resulted from these primary changes would have led to other adaptive developments. A longer juvenile phase could have led to an increase in height because the major period of bone growth and particularly bone lengthening occurs at this time. And this may have had further consequences; an arboreal life style could have been progressively restricted because a taller, long-boned hominid would not have been quite so agile in the trees.

A bigger brain would have needed more time to develop in the uterus before it could function outside; thus we can speculate that there would have been an extension of the gestation period too. A bigger brain would have led to a bigger head size, but head size couldn't increase beyond the limits imposed by the constrictions of the birthing process. So a maximum gestation period would have been reached. If the gestation period became any longer and the foetus any larger, the birthing process would no longer be feasible. (As it is, humans appear to have more difficulty giving birth than other animals.) But with such a large and complex brain, this available period of gestation would not have been long enough to complete the brain's development. In the higher hominids, this would have led to a situation in which the infant's brain would not have been fully functional at birth. At birth, the infant would have still been in a prenatal state that would necessitate an increased level of maternal care. This would have created behavioural and physiological pressures. These pressures would have been present to some degree in the early, smaller hominids but they would have been increasingly significant as body and brain size increased.

The period of pregnancy, the size of the baby, its helplessness at birth, and for an extending period of time thereafter, would have put many demands on the mother but in effect, on the whole troupe/social group as well. Solutions to these pressures may have been varied but in one or more lines of hominids, bipedalism may have been an answer. Standing and walking upright may have been a response to dealing with the extended postnatal stage. Uniquely in the later hominids and especially in humans, the foetal stage in effect continues after birth. A large helpless baby needs to be looked after and protected. It would have become increasingly difficult for the mother to carry this helpless and growing child around in the trees, so a sustained period on the forest floor may have been a likely option. Because human babies are helpless for so much longer than the infants of closely related primates they became

27

proportionately more demanding for longer. Solutions to this enforced ground-dwelling phase need not have included bipedalism but this mode of locomotion does have some advantages in these circumstances. And almost certainly it was a necessary precursor to maximum brain expansion. It is not the only solution however; chimpanzees sometimes walk rather inefficiently on two legs part of the time, so partial bipedalism or continued four-legged locomotion would have been the preference of some primates. But becoming upright could have been an efficient solution to a burgeoning problem. Walking on two legs frees up both arms to carry the infant. It increases the visual range, a helpful adaptation on the less secure forest floor, and it allows both hands (in the case of the mother one hand) to be used for foraging.

Bipedalism may have even been a response to a flooded forest floor. The Congo River basin, the home to our closest living relatives, has areas that become seasonally inundated with water. The Amazon, the Congo's sister system in South America, has a total forested area the size of Great Britain that is flooded for six months of the year. Could such a habitat have provided an impetus towards walking upright? Elaine Morgan in her thought provoking book 'Scars of Evolution' sets out the case for an aquatic phase in the evolution of hominids. The so-called 'aquatic ape theory' does have some elements that are intriguing, though more for the questions it raises rather than its ultimate conclusions. For instance, Morgan identifies many of the structural and functional problems of walking with a perpendicular spine (spinal compression, back pain, inguinal hernia, varicose veins and haemorrhoids) that would be partially ameliorated by spending large amounts of time wading in water. She then goes on to point out that:

'Walking erect in flooded terrain was less an option than a necessity. The behavioural reward – being able to walk and breathe at the same time – was instantly available. And most of the disadvantages of bipedalism were cancelled out. Erect posture imposes no strain on the spine under conditions of head-out immersion in water. There is no added weight on the lumbar vertebrae. The discs are not vertically compressed. (An astronaut in zero gravity gains an inch in height in the first days in space, and immersion in water is the nearest thing to zero gravity on planet Earth.)
In water, walking on two legs incurs no more danger of tripping over and crashing to the ground than walking on four. There is no distension of the veins because immersion in water prevents the blood from pooling in the lower limbs. Water thus seems to be the only element in which bipedalism for the beginner may have been at the same time compulsory and relatively free of unwelcome physical consequences.'

If the pressures we alluded to earlier were forcing our hominid lineage to spend more time on the forest floor and that forest floor was flooded for part of the year, perhaps there was a need to wade through water to access the fruit trees within their territory. But could this provide a pressure that resulted in bipedalism? We feel it is unlikely, though undoubtedly an upright stance would be of benefit in these circumstances. The theory perhaps makes more sense than the orthodox view that tries to bamboozle us into believing that, several million years ago, a population of apes or early hominids, living on the savannah, chose to stumble around on two limbs instead of running easily, like chimps and baboons, on four. Elaine Morgan again asks could they have:

'... *stood up, with their unmodified pelves, their inappropriate single-arched spines, their absurdly under muscled thighs and buttocks, and their heads stuck on at the wrong angle, and doggedly shuffled along on the sides of their long-toed, ill adapted feet.*'

And the reasons proposed for this unlikely transition – that we stood up to hunt meat, or pick grass seeds, or to carry food back to our families, or to minimise the area of our bodies we exposed to the sun – do not stand any serious scrutiny. The two free arms would have been useful to carry food and a large and helpless infant, and the erect posture may have helped to spot savannah predators, but these advantages are likely to have been a 'secondary benefit' not the driving force. We also know that bipedalism became a specialised feature of hominids not in the later stages of their evolution but as far back as four million years ago. The 'Lucy' skeleton, one of the best known species of early hominid, Australopithecus afarensis, has been characterised as a 'bipedal chimpanzee' and recent work in Kenya has unearthed an even earlier species of bipedal hominid known as Australopithecus anamensis. The structure of the pelvis and the knee joints of Lucy and her cousins show that they were upright walkers, but the length of their limbs and the structure of their hands and feet also attest to their arboreal nature. These early hominids were perhaps less adept in the trees than present day apes, and less efficient bipeds than Linford Christie, yet this 'half way house' was successful for this adaptation endured for some two million years. These early hominids were living in the forest, eating a diet probably not dissimilar to that of chimpanzees and bonobos, and the elements of that diet were found primarily in the forest canopy.

The veracity of this scenario has been strengthened by the very latest finds at Kapsomin in Kenya's Baringo district. In December 2000, the Kenya Palaeontology Expedition (KPE) reported the discovery of what is almost certainly a new species of hominid. The excavating team, that included Martin Pickford from the KPE and Brigitte Senut from the Museum of Natural History in Paris, unearthed thirteen fossils belonging to at least five individuals, both male and female. These finds represent a hominid that is far older than any other previous discovery. It has been tentatively dated at least 6 million years old, which means it would be some one and a half million years older than Australopithecus anamensis, the previous most ancient hominid, and older than 'Lucy' too.

This new hominid has been popularly dubbed 'Millennium Man' but its Linnean name Orrorin tugenensis attests to its discovery in the Tugen Hills. What is most exciting about the find, however, is the creature's structure. An almost perfectly fossilised left femur shows Millennium Man had strong back legs that enabled it to walk upright, giving it hominid characteristics that relate it directly to the bipedal lineages. The postcranial evidence also suggests that Orrorin tugenensis was already adapted to habitual or perhaps even obligate bipedalism when on the ground. A thick right humerus bone from the upper arm points to its considerable tree-climbing skills and the length of the fossil bones show the creature was about the size of a modern chimpanzee. The teeth and jaw structure suggest a similar dentition to modern man. It had small canines and full molars that indicate that it would have eaten a diet of mainly fruit and vegetables.

Preliminary analyses therefore indicates that the hominid was about the size of a chimpanzee, an agile climber, that it walked on two legs when on the ground and ate a diet of fruit and vegetable matter. Although it is easy to fall into the trap of extrapolating from minimal fossil finds, this outstanding discovery does strongly support our hypothesis that bipedalism developed in the forest in an animal that was an agile tree climber and ate a diet of primarily fruit and leaves.

Further studies of Orang-utans by Dr Robin Crompton of Liverpool University also indicate that the first stages of bipedalism developed in the trees. He has looked in detail at the ways these large primates move along branches and has noted similarities between their gait and fully functioning bipeds. This work, together with the associated evidence we have noted, strongly suggests that the initial stages towards bipedalism arose in the trees, and the other pressures that we have identified encouraged the process on towards a fully functional two-legged gait.

Pressures acting on the infant stage may have further propelled bipedalism. Apes can climb from a very young age but human infants are very different. After a long period of being completely dependent on their mothers, there is a period in which they are more independent but not developed enough to climb on their own. This would have provided a window of opportunity for the infant to experiment with different modes of locomotion. It may have been of great benefit for the young child to develop an efficient way of getting around during the increasingly long period before it had the strength and dexterity to climb into the trees. And successful children would survive better and pass their traits to their children. If we look at the way children develop their motor skills today, we can see that they are not strong or balanced enough to climb until well past the age they struggle to become upright. In the context of our arboreal origins it is worth pointing out that even today, as adults, we still possess impressive climbing abilities. In the rainforests people regularly climb trees to gather honey and fruit. In 'developed' societies too many people enjoy climbing as a sport and gymnasts, particularly on the parallel bars, display superb arboreal skills.

The point we would like to emphasise in all this is that bipedalism developed within the forest and the primary instigating factor was the changes brought about by the biochemistry of the diet. The developmental window would have become longer as the juvenile period extended. Initially there may have been an enforced two or three months on the ground but as the biochemical changes took effect this may have been extended to a period of one, two or even three years. The hominid's arboreal features would have been retained as much as possible but these would have been progressively constrained by the hominid's increasing size. We can see then that this change is not coming through the normal DNA selection route but via physiological changes brought about by the action of the chemicals contained in the diet.

HAIRLESSNESS AND VITAMIN D ~ψ~

It is all too easy to assume a simplistic picture of human evolution. The story goes that we separated from our nearest relatives, the chimps and bonobos, somewhere around seven million years ago and this line led via various strands of hominids to the Neanderthals and us. The true scenario was in all probability much more complex and fragmented.

Various lineages would have branched off, moved away from the forest, migrated, settled along coastlines, lived and eventually died out. There may have been back-crossings, for different lineages would certainly have been genetically compatible for a long time. Llamas and camels, separated by five million years of evolution, are still able to interbreed – these hominid lines would have been much closer. Some evidence of Human/Neanderthal interbreeding has been suggested from recent finds in Southern Spain. Thus we can safely speculate that different races of hominids would have changed physiologically and then by crossbreeding returned to nearer the hypothetical source population. A highly complicated weaving of strains could have occurred which would cloud and confuse the picture of human ancestry. But at least one lineage would have remained in the forest, after all this was the safest and most nurturing habitat to dwell in. These populations would continue to be subjected to the physical and biochemical pressures imposed by this hot-house environment.

Inevitably over time, further changes would have occurred and one may have been a result of the necessity to increase the efficiency of vitamin D absorption.

There has been a great deal of confusion about our hairlessness – why should *Homo sapiens* be so lacking in body hair? Even Charles Darwin could see no advantage of nakedness to man. The 'Father of Evolution' concluded that 'our bodies could not have been divested of hair through natural selection' but somewhere along the line this is presumably what indeed did happen.

Ideas to explain our nakedness have included:– deterrence to parasites, a cooling system, a lure to increase our attractiveness to the opposite sex, and a response to living mainly in water. None of these explanations are particularly convincing. Hairlessness has not deterred ticks and leeches. In hot scenarios, fur actually protects against the sun and in the open savannah, where this is supposed to have taken place, fur would be needed for warmth at night. And no other primate has lost its hair, certainly not as a sexual adornment. The water theory may have more substance. Parallels have been drawn with large aquatic mammals that have developed layers of blubber under a smooth, hairless or closely furred skin. We do indeed have fat tissue beneath our skin but fat just wouldn't have been an issue for a hominid living on a diet of mainly fruit. It is difficult to get fat on a diet of wild fruit. Most of the enlargement of the human blubber layer is a symptom of overeating, which is increasingly easy on a diet of particularly refined carbohydrates.

Did our ancestors then really spend aeons living in water? Is it possible that, sometime between leaving the forest and reaching New York, humans went through a phase of living almost entirely in water? How did they deal with powerful predators like crocodiles? A naked ape in the water would have been extremely vulnerable to such ferocious creatures. (Even today salt-water crocodiles occasionally terrorise villagers living near estuarine swampland in Southeast Asia.) Unlike other mammals, unless we actually learn to swim, we can easily drown when we fall in water – we do not seem to have any instinctive swimming ability. This would be a surprising oversight for an animal with an extended aquatic phase. And why does our skin go all wrinkly when we sit in the bath or swim in the sea? We just don't appear to be well enough adapted to be an aquatic species, yet anomalies exist that have not been satisfactorily explained. Why do we have far more sebaceous glands than our nearest primate relatives do? And why do we sweat when most animals pant to reduce body heat? A new explanation is required. All current theories are seriously flawed.

If water was an element in our development it must have only been a contributory factor. It is just tenable that in the evolutionary history of mankind there was a period in which water created some adaptive pressure. Along the huge river systems of the Congo and the Amazon the usual boundary between water and forest is blurred due to seasonal inundation. If water was a factor in establishing bipedalism and hairlessness, it may have been that during the wet seasons water came into the forest, not that mankind's ancestors left the forest to live aquatically. It is interesting to note that in the Congo gorillas spend time foraging for tasty shoots of water plants. They spend much time in the water but haven't become hairless. Could this sort of lifestyle really be the reason humans became hairless? It is doubtful.

Vitamin D is important to our health; without it we cannot absorb and assimilate calcium. A deficiency of this vitamin can cause bone disorders like rickets and a weakness of bones certainly would be a major disadvantage when living even a partially arboreal lifestyle. Vitamin D is unusual as it is in short supply in an exclusively plant-based diet and our bodies do not absorb it very efficiently from animal products either. But cells in the skin can manufacture it when the skin is exposed to sunlight. A hairy primate gathering fruit in the tops of the canopy would be subjected to enough light to maintain vitamin D production but a larger primate forced to spend more time on the dark forest floor may have needed to increase the efficiency of the production mechanism. It is possible that hair loss resulted as a response to this pressure, for the sunlight that filtered down to the forest floor would need to be utilised to the maximum. It is vaguely possible however that the continued arboreal life style, with at least some time spent in the upper canopy searching for fruit, could have exposed the top of the head to deleterious effects of direct sunlight. The retention of hair on the scalp could thus provide in these circumstances a positive benefit. Not a strongly convincing argument but perhaps it is a little more likely than another suggestion that we have played around with – that head hair provides a convenient hand hold on an otherwise slippery parent.

More seriously, there may also be a steroid factor here. Vitamin D is chemically very similar to a steroid. Did the steroid-suppressing chemicals in the fruit inhibit the activity of vitamin D, making the need for a more efficient absorption mechanism all the more vital? Perhaps this extra selection pressure tipped the balance in favour of hairlessness. An imbalance of steroids could also explain the anomaly of our over-active sebaceous glands. We know that if males are castrated the activity of the sebaceous glands (and acne) decreases. So it is possible that today our sebaceous glands are working overtime in response to a heightened internal steroid environment. When we were naked in the forest our internal steroid activity would have been lower due to the continuous chemical effect of the steroid suppressing chemicals assimilated via our diet.

Losing hair in the tropical forest environment would not have incurred any major penalties. And while it is extremely unlikely that hairlessness came about as a mechanism for sexual attraction, the sexes may have come to like the look of each other this way. Hair loss may have even had some other advantages such as enhancing radiant heat loss though, as other primates in such environments retained their hair, this may have been a minimal or secondary benefit. Naked skin would also increase the efficiency of sweating as a cooling mechanism. Most animals do not sweat anything like as much as humans but regulate their temperature by panting. The problem with sweating is that it involves substantial loss of water but this would not have been a serious disadvantage in the humid forest, particularly for an animal that ate a fruit/leaf diet that comprised 80% to 90% water. But transfer this trait to the savannah and it does become a serious drawback. Savannah hominids living under the glare of the sun would have to cope with the dangers of dehydration. Certainly sweating as a way of cooling the body, like so many other traits that we regard as uniquely human, could not have evolved in this habitat. All in all, the savannah model is looking increasingly untenable. It has even been discounted now by previous stalwart supporters like paleoanthropologist Phillip Tobias.

It is easy to assume that there was just one lineage that turned into an animal as advanced as a human. Though this may have been the case, it is more likely that there were many branches and many different adaptive solutions to the problems posed by all the various environmental pressures. But, in the one particular lineage that happened to survive, all these solutions that we have been discussing came together. There may have been human-like hominids that were not naked in the forest. There may have been a human-like lineage that didn't develop such efficient bipedalism, but the one lineage that brought all these elements together is the one that survived.

BIG BRAINS ~ψ~

The tropical fruit diet of our ancestors, as we have constantly reiterated, had a marked effect on our anatomical and physiological development. The most significant effect however was reserved for our brains. We believe the biochemistry of a fruit diet became the necessary foundation for a chemical drive that turned relatively big-brained primates into great apes and extremely large-brained humans. The chemicals within our ancestor's fruit diet fuelled gradual change and were essential to maintain this change, but what emerged from this was an internal feedback loop that at some point compounded the effect. This second wave of chemical change was an internal mechanism. Maybe neural development in great apes has gone as far as it can without this internal mechanism really taking off. It must be remembered that gorillas, chimpanzees and humans are genetically extremely close and, of the three, humans are genetically closer to chimpanzees than chimps are to gorillas. Why then are there not three species of African great apes? Why are we so genetically close, yet so different?

As natural selection cannot fully explain the difference, another mechanism must be responsible. We suggest that this internal biochemical mechanism is the crucial missing part of the jigsaw that propelled humans into a completely different league as far as brain size is concerned. Out of the millions of species that have arisen on this planet, humans display unique traits. Does this not imply there has been a unique process going on? It is encouraging to read that Professor Colin Groves, author of one of the most respected books on primate and human evolution, has also suggested that brain growth may not have been selected, as such, but happened fortuitously as a result of the changing tempo of our patterns of growth. In a personal communication he confirmed that he still thinks it is likely that human brain expansion occurred as an 'epiphenonomenon of neotony'. We will be exploring the detail of the fundamental biochemical pathways and biofeedback mechanisms that affected these patterns of growth in the next chapter.

The ideas of Professor Groves are supported by the fact that bonobos possess great latent intelligence. This intelligence arises from their big brains, not vice-versa. Bonobos do not use tools in the forest, at least to the extent that chimps do, but when transferred to a different environment, i.e. when kept in captivity, bonobos display a greater capacity for tool use (and possess higher levels of cognitive skills) than chimps. In their natural environment, bonobos do not actually need to use tools, as everything they require for survival is readily available. The implication is that bonobos have a greater capability for intelligence than they need, and thus intelligence is a result of their big brains rather that big brains arising out of the need to develop such things as tool use. As Colin Groves believes – big brains are a biproduct of some evolutionary mechanism.

CAST OUT OF THE GARDEN ~ψ~

At some stage humans left the forest. At different phases along the hominid evolutionary way individuals, groups and tribes would have left the forest ecosystem to disperse or find new territory. Pressures like climate change and accompanying forest contraction may have also be significant. If they were adapted to eating fruit and the fruit was plentiful, there would be no overt pressure to leave and indeed chimps, bonobos and gorillas are still there, eating a mainly vegetarian diet. We do know however that a cooling climate around the critical time would have resulted in forest shrinkage. Waves of hominid species and even the race that came to be known as Neanderthals may have been forced out of the prime habitat while the hypothetical lineage that we have been following (and it would have been the best adapted one) stayed in the forest. But even this lineage left perhaps somewhere in the region of 200,000 years ago. It is accepted that around this crucial juncture there was a reduction in global temperatures and rainfall, which led to a fragmentation of the forest. If a big central band of forest split, the populations within it would be fragmented too. If then these smaller forests declined from climactic or even celestial pressures, the populations within would be rendered homeless.

Something evidently forced the whole population out. Perhaps a catastrophic event like a meteor impact tipped the balance. We are now beginning to realise that these catastrophic events were far more numerous than previously thought. In 'Evolutionary Catastrophes' Vincent Courtillot states; 'over the past 300 million years our planet has been battered by at least seven major ecological catastrophes'. An extraterrestrial impact or violent vulcanism may have knocked out the fruit producing capacity of the remaining forest for long enough to have forced an intelligent human to look elsewhere for sustenance.

From mitochondrial DNA analysis we know that at some time in the past our population was extremely small. The billions of humans on our planet are all descended from perhaps as few as 5000 individuals. This suggested bottleneck could have been as a result of an ecological disaster that ravaged population levels, though it is likely that populations in the forest were always small and inbred. Large numbers of humans did not appear until very much later in our history. Bonobo and chimpanzee populations today are relatively small and the breeding rate low, despite the excessive amount of sexual activity particularly displayed by the bonobos. We will be looking at why a predominantly fruit diet may have been a contributory factor to this low level of fertility and slow breeding rates later.

Leaving the forest was a highly significant change. The adaptations we have looked at in previous sections (a big brain, an extended juvenile period, a loss of hair, an efficient two-legged gait and a partial ground-dwelling habit) all stem from the chemical effects of a fruit-based diet. When this drip feed of fruit chemicals was interrupted, some change was inevitable. It would almost certainly be deleterious, as it would to any creature ousted from its natural habitat. A naked and placid hominid, from an environment in which everything was provided, was suddenly exposed to new and harsh conditions. The biochemistry that its optimum neural function depended upon was gone and it had to fend for itself in extremely different circumstances. But a big brain and intelligent adaptability would have given our ancestors a better chance of meeting the challenges of a different environment. Humans have a physiological adaptability too – we are able to survive on many different types of diet ranging from totally plant-based to almost exclusively animal. But despite this, when the fruit part of the equation was lost, neural and physiological function must have been negatively affected. Without the biochemistry that was the foundation for our unique development, changes would have been rapid and irreversible. These changes could have even led to the initiation of an unstoppable negative feedback mechanism: even if the forest became habitable again and humans returned to their former way of life, there would have been no guarantee that the brain

36

expansion process would have restarted. During the intervening period, without the forest biochemistry, structural changes would have been initiated; different biochemistry builds different structures, which leads to the emergence of different traits.

We have paid a major price for this change of lifestyle. Although, from the classic survival/competition perspective we are more successful now than we have ever been (and there are some six billion of us), we suspect that the brains of our close ancestors living in the forest had greater potential than ours do today. We may have slipped away from an opportunity for even greater and possibly more balanced brain development which would have given us much greater function and a more benign sense of well being than we possess today.

The forest legacy has left us with a colossal piece of equipment between our ears – give us a problem and we will solve it – but without the foundation of the optimum chemistry provided by a fruit-based diet we may no longer be able to access its optimum performance.

AGGRESSION ~ψ~

A classic and rather romanticised image of a placid but physically capable and strong primate/hominid living a carefree life in the forest may not be too far from the truth. It is highly likely that the inherent physiological and psychological state of our ancestors would have been one of underlying ease and contentment. Pumping chemicals from the forest fruits into the hominid system would have limited aggressive behaviour. Steroids are linked to aggression and, as we have seen, many of the chemicals found in fruit suppress steroids. They dampen down the effects of testosterone in particular.

The role of androgens, such as testosterone, in precipitating aggressive behaviour is clearly displayed in hyaenas. In these extraordinary animals females are not only dominant but also highly aggressive – young females regularly kill their siblings. This aggression has been linked to high levels of testosterone. In fact the females have more testosterone than the males and even possess pseudo male genitalia.

Internal steroid activity would have been altered when humans left the forest and fruit was lost as the major part of the diet. Not only would steroid suppression have been lifted but also eating different foods instead would have made matters even worse. Meat contains steroids, thus a change to a carnivorous diet would have caused a major biochemical upset – not only would the steroid suppressing elements in the diet have been lost but also extra steroids would be taken in. Recent research has found that eating animal fats increases levels of testosterone: A double whammy is the current phrase that springs to mind.

It is sobering to note that humans killed in excess of 100 million fellow humans in the twentieth century alone. Today we live surrounded by mental, emotional and physical torture. There is cruelty to others and to the animals we live with, and we are rapidly destroying the planet that sustains us. Is it possible that this sickness (and it really is a sickness) stems from an imbalance in our biochemistry initiated by the loss of steroid suppression all those years ago? Without doubt, something has gone wrong. In his famous book 'The Ghost in the Machine' Arthur Koestler drew similar conclusions:

'When one contemplates the streak of insanity running through human history, it appears that Homo sapiens is a biological freak, the result of some remarkable mistake in the evolutionary process. The ancient doctrine of original sin, variants of which occur independently in the mythologies of diverse cultures, could be a reflection of man's awareness of his own inadequacy, of the intuitive hunch that somewhere along the line of his ascent something has gone wrong. (p267)

To put it vulgarly, we are led to suspect that there is somewhere a screw loose in the human mind. We ought to give serious consideration that somewhere along the line something has gone seriously wrong with the evolution of the nervous system of Homo sapiens. We know that evolution can lead into a blind alley, and we also know that the evolution of the human brain was an unprecedentedly rapid, almost explosive, process. Let us note as a possible hypothesis that the delusional streak which runs through our history may be an endemic form of paranoia, built into the wiring circuits of the human brain.' (p239)

Would it not be wonderful to identify the mechanism of our insanity, for then we could do something about it? Real healing and restoration of a balanced consciousness may be realistic but initially we need to recognise the depths and significance of the problem.

WHY DID OUR BRAIN STOP EXPANDING? ~ψ~

In the forest, the human brain was expanding and expanding at a phenomenal rate. Sometime at around 200,000 to 150,000 years ago this process came to an end. The brain stopped expanding and started to shrink. This key point in our evolutionary journey has been noted but rarely addressed, and its significance comprehensively ignored.

Christopher Ruff, of John Hopkins University, and his colleagues thoroughly analysed the fossil record to determine the evolving body mass and brain size of the Homo species leading up to us. The results show that the assumption of a straight progression from a pea-brained ancestor to the ultra-brainy modern *Homo sapiens* is decidedly shaky. Hominid brains appear to have remained fairly constant in size for a long period from some 1.8 million years ago until about 600,000 years ago. But then, from 600,000 years to 150,000 before the present, fossils show that the cranial capacity of our ancestors skyrocketed. Brain mass peaked at about 1,440 grams. Since then brain mass has declined to the 1,300 grams that is typical today.

Of course brain size alone does not tell the whole story. Brain size also correlates with body size and the peak of brain size roughly corresponds to the peak in archaic *Homo sapiens* body size (the Neanderthals). The decline in size of the body in *Homo sapiens sapiens* (modern humans get two 'wises' in our name, but do we really deserve it?) over the past 50,000 years has raised our ratio of brain to body size to just above Neanderthal levels. Yet we have done this by shrinking our bodies to a greater extent than our brains have shrunk. There is some evidence that our brains are still shrinking, and may have done so over the last 10,000 years by as much as 5%.

This very recent period of brain shrinkage coincides with a major dietary change, for it was around this period that cereals and grain came to the fore. Cereal grains may be the foundation of our diet today and responsible for the huge explosion in our numbers but they may not be the best of foods for optimum function. Indeed studies of skeletons from early agricultural societies show ill health accompanies the initial transition to eating more grains and cereals. Skeletons dug up from the East Coast of America, dating from around 1000 AD, the era when Native Americans switched to corn-based agriculture, are smaller than earlier skeletons. Studies of skeletons from other societies undergoing this transition show signs of deficiencies such as anaemia. Clark Larsen, the physical anthropologist who studied the East Coast skeletons has stated that 'just about anywhere that this transition to cereals occurs, health declines'.

It is thought that humans from such agrarian societies were lucky to live beyond thirty years. In contrast forest apes, such as chimpanzees, can live for some sixty years. We can reasonably assume that humans in the forest lived easily as long if not longer. Furthermore, if man in the forest was as long lived or even longer lived than chimps, it would provide a strong argument for the notion that this was both the most natural and most suitable place, particularly in terms of diet, for a human to live.

ANCESTRAL DIETS ~ψ~

If the evolution of the unique human system were somehow linked with our ancestral diet we would expect the human system still to be best adapted to something approaching this.

While there is continued debate on this subject, few dissent from the view that there is an increasing problem with the food we are eating in our sophisticated time-stressed modern world. In just one six-week period, newspaper headlines in the United Kingdom announced: 'World alert over cancer chemical in cooked food' (Daily Telegraph May 18th, 2002); 'Children at risk from the junk food time bomb' (Daily Mail May 31st, 2002); and 'Anti-social conduct may be linked to diet, says study' (Guardian, June 26, 2002). This is a small sample of worries arising from recent research. Today, we are told we risk diabetes, heart disease and cancers from eating the 'wrong sort of food'. Weight problems caused by an addiction to high fat and high sugar convenience foods, or simply an ignorance of the alternatives, carry the risk of these and other diseases manifesting in later life. One in ten children under four is now classified as obese and health problems resulting from being overweight costs Britain some two billion pounds a year. It has been estimated that, if we continue eating a 'junk food' diet, in forty years time half the population will be obese. Furthermore specialists also fear that anaemia due to poor nutrition in early life can have long-lasting effects on a child's mental development and learning ability.

Although longevity has increased over the last few centuries, many folk live the last years of their lives with the fear of disease, if not the actuality of it, but old age and disease do not necessarily go together. In the remote Andean highlands of Ecuador, there are communities of people who it is claimed live for 140 years or more and who remain agile and lucid right to the end. Death from heart disease and cancer is unknown in these high mountain valleys but rife in nearby towns. David Davies, who has made a study of these 'Centenarians of the Andes', found that the people who have the best chance of a healthy old age are those that actively use their minds and bodies, even towards the end of their life span. He looked at many elements of their life and environment from genetic factors to the tranquillity and lack of stress in their way of life. The folk who lived longest were found amongst those that lived on a subsistence diet, which was low in calories and animal fat. Typically, the main meal of the day was eaten in the early evening and was made up of very small wild potatoes, yukka, cottage cheese and maize or bean gruel. Melons were eaten for dessert. Sometimes green vegetables, cabbage, marrow, pumpkins were added to the menu and sweet corn cobs were often taken to work for lunch. The people working in the fields ate fruit throughout the day. The climate is ideal for citrus fruits, and many other 'hedgerow' fruits such as mora (like a blackberry), guava and naranjhuila are abundant too. Meat was only eaten rarely, a type of cottage cheese was made from goat or cow milk, and eggs were eaten raw or almost raw.

Though these people are very healthy and extremely long-lived we mustn't necessarily jump to the conclusion that this diet is perfect for the human system – their diet is restricted by the environment they live in. However, if we look at other communities of long-lived folk the parallels are striking. The Hunzas of north-east Kashmir also live in mountainous regions and have a diet that includes wheat, barley, buckwheat, beans, chick-peas, lentils, sprouted pulses, marrows, pumpkins, cottage cheese and fruit – the famous Hunza apricots and wild mulberries. Meat is again only eaten rarely and, because fuel is in short supply, when food is cooked it is usually steamed; a method of cooking that is the least damaging to the chemical nutrients in the food. Hunzukut males, like the people in the Andean Highlands, are also reported to live to 140 years of age. So, we must conclude that these diets are, at the very least, much more suitable than the ones we depend on in the affluent industrialised countries.

There seems to be no definitive study that has so far convinced society as a whole that nutritionally we are barking up the wrong tree (or at least not picking from the right one). But there are many scraps of information that support the thesis that a more natural diet is the most beneficial option. Lymphocyte production and hence resistance to illness is boosted by consuming the nutrients that occur in optimal proportions and quantities in uncooked vegetables. There are also a huge number of cases in which raw food, particularly fruit and vegetable juices, has seemingly cured a wide range of illnesses. Migraines, skin complaints, tuberculosis, mental disorders, heart disease, cancers and a host of other diseases have responded favourably to a diet rich in raw food. There are clinics, foundations and institutions throughout the world that offer therapies based on 'living nutrition'. Such diets are much closer to our ancestral diets than the chips, pies and biscuits that adorn most of our supermarket shelves.

As with all organisms, hominids in the course of evolution were locked into the biological matrix of their environment. Whether our diet consisted of insects, fruit or meat it was all biologically active material. Some primates today eat a bit more of this or that – much coverage has been given recently to meat-eating chimps but this comprises a relatively small percentage of their diet. Despite their skill in capturing live prey, chimpanzees actually obtain about 94% of their annual diet from plants, primarily ripe fruits. Primate biochemistry is largely based on plants and a plant-based diet is what hominids were eating during their evolutionary development. A pictoral representation of an early human living in the forest, lounging around eating fruit, may be more accurate than one in which he is dressed in animal skins, spear in hand, on the hostile open plains.

The lack of plant material in the fossil record has led, according to Richard Leakey, to an over emphasis on meat eating as a component of the early hominids' life. He also finds some of the recent work on tooth analysis 'surprising'. The teeth of Australopithecus robustus fall into the fruit-eating category. The patterns of wear and the small scratches left on the enamel appear very similar to that of the forest dwelling chimpanzees, yet here was a hominid which was supposed to live on the plains in an era when the climate was dry and the vegetation mainly grass. The examples of Ramapithecus teeth that have been similarly analysed show exactly the same pattern, and the teeth of Homo habilis, the first creature to be awarded Homo status, also has smooth enamel typical of a chimpanzee. This evidence is extremely relevant. All the early hominids and their great ape cousins were mainly fruit eaters. The teeth of Homo erectus suggest a more omnivorous diet. The enamel from their teeth show scratches and scars that are compatible with grit damage possibly from consuming bulbs and tubers. As a response to a cooling climate and a contraction of the forest, did this species widen its diet to adapt to a new environment? Some forest would have remained intact along the wetter river systems. Chimpanzees and gorillas survived there along with, we suspect, another hominid whose teeth were very well adapted to fruit eating – Homo sapiens.

Primates, given a choice, will select fruit in preference to any other food. Fruit is a rich, nutritious and easily digestible food. If it is available, this is what all the great apes prefer to eat. However other foods are eaten regularly. Our nearest relative, the bonobo, eats between 60% and 95% fruit depending on the fruit productivity of its specific habitat. The rest of its diet comprises mostly shoots and herbs and a small amount of insects, eggs and the occasional small mammal. Fallback foods like bark may also be eaten in times of fruit scarcity.

What humans in the forest ate is of course unknown but it is likely that they would have eaten a similar balance of foodstuffs. They would not have been purely 'vegetarians'. Even figs (perhaps the most preferred food) contain a small amount of insect matter as their pollination mechanism results in eggs and larvae of small wasp species remaining in the fruit. These insects may have served as an important source of essential micronutrients such as vitamin B12 as well as providing a little extra protein.

As they were the most highly intelligent animals in the forest and fruit was the best food, it is likely that humans developed strategies to maintain a high percentage of fruit all the year round. Being efficient bipeds would have given them the potential to travel easily between widely separated fruit sources. The quest for distant fruit trees may have even honed their bipedal adaptation. The larger arboreal primates are known to travel on the ground between distant fruit trees, as it is more efficient than travelling in the trees. Archaic humans, being better-adapted bipeds than apes, would have found this way of life much easier.

HUMANS ARE BY NATURE FRUGIVOROUS ~ψ~

There has been much study and even more speculation about what sort of diet our teeth and guts are best designed for. From the type of dentition, gut length and toxicity of foods like meat, a very strong case can be built for Homo sapiens being designed to eat and process a largely fruit-based diet. The brain's requirement for food and the gut's requirement for energy, optimal acid/alkali balance and the structure of the intestines all point to a frugivorous diet. A shift to fruit specialisation answers all the problems and anomalies that have spawned countless conflicting theories.

Katherine Milton, Professor of Anthropology at Berkeley University, California, has carried out important work on diet and primate evolution. Her research has led her to believe that 'the strategies early primates adopted to cope with the dietary challenges of the arboreal environment profoundly influenced their evolutionary trajectory'. This has a great significance for us today for the foods eaten by humans now bear little resemblance to the plant-based diets anthropoids have favoured since their emergence. She believes these findings shed light on many of the health problems that are common, especially in our industrially advanced nations. Could they be, at least in part, due to a mismatch between the diets we now eat and those to which our bodies became adapted over millions of years?

The plant-based food available in the forest canopy comprises fruit and leaves but subsisting on this diet poses some challenges for any animal living here. For a start, it is high in fibre that is not only difficult to break down and hence digest but also takes up space in the gut that may otherwise be filled with more nutritious foods. Many plant foods also lack one or more essential nutrients such as amino acids, so animals that depend on plants for meeting their daily nutritional requirements must seek out a variety of complimentary food sources. Fruit is usually the food of preference for it is rich in easily digested forms of carbohydrate and relatively low in fibre, but its protein content is low too (their seeds may be protein rich however). Leaves offer a higher protein content but they are lower in nutrients and contain much more fibre. Balancing these constraints have led to different strategies that are reflected in behaviour and physiology. Colobine monkeys have compartmentalised stomachs (a system analogous to ruminants) that allows fibre to be fermented and hence processed very efficiently, but humans and most other primates pass fibre largely unchanged through their digestive systems. Some fibre can be broken down in the hind-gut of these latter species but the process is not as efficient as that in the Colobus.

Milton's research focused on two contrasting species of South American primates howler and spider monkeys. These two species are about the same size and weight as each other and live in the same environment, eating plant-based foods, yet they are very different. Howler monkeys have a large colon and the food passes through its digestive system slowly, whereas spider monkeys have a small colon through which food passes more quickly. These physiological differences relate to dietary specialisation. The foundation of the howler's diet is young leaves: 48% of their diet is leaves, with 42% fruit and 10% flowers. The spider monkey's diet comprises 72% fruit, 22% leaves and 6% flowers. Another fundamental difference is that, although these animals are the same size, the brains of spider monkeys are twice the size of howlers. Very significantly, Milton comments that 'the spider monkeys in Panama seemed 'smarter' than the howlers – almost human'. This is something we have commented on before: big brains and a diet high in fruit appear to go together. Why should this be so? Could this brain enlargement result from the need to memorise the location of productive fruit trees, as some have suggested, or did, as we propose, elements within the fruit itself fuel this change more directly? Animals such as squirrels, and even birds like jays, memorise the locations of stored food most efficiently without an overlarge brain thus it seems that something else must be responsible.

Although Milton has concluded that it is quite difficult for primates to obtain adequate nutrition in the canopy, she observed that spider monkeys consume ripe fruits for most of the year, eating only a small amount of leaves. Bonobos also appear to find enough food to eat easily, for much of their time is spent in other 'social' activities. Thus, being a fruit-eating forest primate appears a very viable option – but one question remains: if fruit is so low in protein, how do these fruit specialists obtain an adequate supply of these essential nutrients? Milton found that spider monkeys pass food through their colons more quickly than leaf-eaters such as howler monkeys. This speed of transit means that spider monkeys have a less efficient extraction process but, as much more food can be processed, it more than makes up. By choosing fruits that are highly digestible and rich in energy, they attain all the calories they need and some of the protein. They then supplement their basic fruit-pulp diet with a very few select young leaves that supply the rest of the protein they require, without an excess of fibre. Of course, by processing so much fruit, a large quantity of chemicals that naturally occur in fruit will also be absorbed. It should also be noted that wild fruit contains a higher percentage of protein than the cultivated fruit that is available to us humans today. It is clear that many wild primates are able to satisfy their daily protein and energy requirements on a diet largely or entirely derived from plants. It is likely that our ancestors in the forest did too.

The wild fruit that we propose was the mainstay of our ancestral diet for the longest and most significant part of our evolutionary history contains more fibre than the fruit we buy today in our shops. Chimpanzees take in about 100 grams of fibre a day compared to about 10 grams that the average western human consumes. At one time it was believed that humans did not possess microbes capable of breaking down fibre. Studies on the digestion of fibre by 24 male college students at Cornell University, however, found that bacteria in their colons proved quite efficient at fermenting the fibre of fruit and vegetables. The microbial populations fermented some three-quarters of the cell wall material, and about 90% of the volatile fatty acids that resulted were delivered to the blood stream. It has been estimated that some present day human populations with a high intake of dietary fibre may derive 10% or more of their required daily energy from volatile fatty acids produced in fermentation.

Furthermore, recent experimental work on human fibre digestion has shown that our gut microflora are very sensitive to different types of dietary fibre. We are very efficient at processing vegetable fibre from dicotyledenous sources (flowering plants like fig trees, carrots and lettuces) but are less so from monocotyledens (grasses and cereals). This provides yet another pointer to the archaic diet of humans as being largely fruit-based and indicates that the grass seed that we eat so much of today in cereals, biscuits and much else is a poor substitute.

The chimpanzee gut is strikingly similar to the human gut in the way it processes fibre. As the fraction of fibre in the diet increases, both humans and chimpanzees increase the rate at which they pass food through the gut. These similarities indicate that when food quality declines both these primates are evolutionarily programmed to respond to this decrease by increasing the rate at which food passes through the digestive tract. And this compensates for the reduced quality of the food available.

It appears that the human system then, like the chimps and bonobos, is designed for a plant-rich fibrous diet. We are not designed for a diet high in carbohydrate and low in fibre or significant quantities of animal protein. Meat eating in man has been, on an evolutionary time scale, a very recent development. It certainly couldn't have influenced the development of our physiology. Though the passage of food through the guts of spider monkeys, chimps and humans is faster than leaf specialists like howlers, it is much slower than carnivores. Meat hanging around in the digestive system is bad news because of its inherent toxicity. The transit time for the passage of food through a carnivore's gut is between 7 and 26 hours while for humans it is between 40 and 60 hours.

Though we do have a shorter colon and a longer small intestine than the great apes (and this has led one camp of researchers to speculate that our intestines are more similar to those of carnivores), these differences are more appropriately explained by a specialist fruit diet, not a carnivorous or grain-based one. Fruit is easier to digest than leaves, tubers and stems, and has a lower fibre content. Thus a specialist fruit eater would not need such a long colon as other apes that have more fibrous bulk to deal with.

Another feature of humans, that is strongly indicative of our vegetarian origins, is our inability to synthesise our own internal vitamin C. This trait is very rare but, where it occurs, the animals concerned (such as guinea pigs) eat a plant-based diet. In these cases ample supplies of the vitamin are available within the food. Vitamin C plays many extremely important roles within the human body. Research seems to be always finding more functions for this 'miracle chemical'. These have been summarised by Dr. Ross Pelton in his book 'Mind Foods and Smart Pills': Vitamin C stimulates the immune system, enabling one to better resist diseases. Terminal cancer patients taking megadoses of vitamin C have been found to live longer. It promotes faster wound healing and reduces the amount of cholesterol in the blood. It is a powerful detoxifier and protects against the destructive power of many pollutants. In addition, it protects the body against heart disease, reduces anxiety, and is a natural antihistamine. A severe deficiency causes scurvy, and eventually death. Increasing intake has been found to increase mental alertness and brain functioning in a variety of ways. Vitamin C is the main antioxidant that circulates in the blood. When available in sufficient quantity, blood carries it around the body, washing over the cells to create a bath of protection. Whenever a free radical turns up, a molecule of vitamin C gives up one of its own electrons to render the free radical ineffective. According to Pelton, this process may take place somewhere between 100,000 and a million times a second, depending on the body's level of metabolism and the amount of vitamin C available. Unfortunately, with each radical decimated, a molecule of vitamin C is lost, so the body rapidly loses its supply of vitamin C.

Vitamin C is a key player in keeping our neural system healthy. The body has a system that operates like a kind of a pump to concentrate vitamin C around our nerves and brain tissue. These tissues have more unsaturated fats than any other organs in the body, making them more vulnerable to attack by free radicals and oxidation. The vitamin C pump removes vitamin C from the blood as it circulates to increase the amount of vitamin C in the cerebrospinal fluid by a factor of ten. The pump then takes the concentrated vitamin C from the cerebrospinal fluid, and concentrates it tenfold again in the nerve cells around the brain and spinal cord. Thus our brain and spinal cord cells are protected against free radical damage by more than a hundred times as much vitamin C as our other body cells.

For such an important chemical, it is extremely odd that we are dependent on vitamin C from outside sources. But how much of it does the body need? Research carried out by the Committee on Animal Nutrition demonstrated that monkeys needed around 55 mg. of vitamin C per kilogram of body weight. When this measure is extrapolated to humans, a 150-pound person would need a daily intake of 3,850 mg. Nutritional science recommends that a human needs 45 mg. each day. This is just enough to prevent scurvy but not enough to keep the body functioning at an optimal level. We would not, and indeed do not, obtain the sort of levels our bodies really need from a diet high in meat and low in vegetables/fruit, but we would from one high in fruit, shoots and leaves. Analysis of wild plant foods eaten by primates shows that many of these foods contain notable amounts of vitamin C. The young leaves and unripe fruit of one species of wild fig was found to contain some of the highest levels ever reported. Our closest living relatives, the great apes, eat a diet that contains between 2 and 6 grams of vitamin C every day. When our ancestors were living in the forest they would have consumed similar amounts.

In contrast, we can and do produce our own vitamin D. This vitamin cannot be obtained from a leaf/fruit based diet but it can from a carnivorous one, thus if we were designed to eat meat we would have less need to synthesise our own. Being able to synthesise vitamin D and not vitamin C is then a strong indication of our true ancestral diet and the one we are really adapted to. Accumulating evidence for meat being an unhealthy food option further strengthens this case. One recent study showed that vegetarians were 24% less likely than non-vegetarians were to die of ischaemic heart disease.

Carbohydates also appear to be problematical when eaten in large amounts. A diet high in carbohydrates, especially refined carbohydrates (cakes, biscuits, pasta, etc), dumps large amounts of glucose rapidly into our bloodstream. This can cause insulin resistance in which the absorption of glucose from the bloodstream is disrupted. This in turn can lead to obesity, adult onset diabetes, hypertension, heart attacks and strokes. It can also lead to an excess of male hormones, which, amongst other effects (aggression), encourages pores in the skin to ooze large amounts of sebum. Acne promoting bacteria thrive on sebum. Up to 60% of 12 year olds and 95% of 18 year olds in modern society suffer from acne, yet it is almost unknown in subsistence societies such as the Kitava islanders in Papua New Guinea and the Ache of the Amazon. The Inuit people of Alaska also used to be free of acne but they began to be affected by these skin complaints after they started to eat processed foods.

The problem with eating highly processed carbohydrates may be further reaching still. If refined cereal consumption results in an excess of male hormones it could have a knock-on effect on the immune system for we know that the thymus gland starts to shrink in response to these hormones at the time of puberty. (More carbohydrates lead to more testosterone which shrinks the thymus gland that is seat of much of our immune response.) Grain products have also been associated with coeliac disease, an auto-immune condition of the gut and some researchers suspect they trigger rheumatoid arthritis too.

It is highly significant that these foods have the ability to alter the quantity or at least the activity of our hormones. It is another example of the way our diet can affect the way our bodies work. It is possible, probable even, that they also affect the way we act and thus moderate our sense of self. If we compare refined carbohydrates with fruit, we can see that fruit has a much lower glycaemic index, which means it is digested more slowly thus avoiding the problems of the 'glucose rush'. The chemicals within fruit also reduce the activity of sex hormones. They thus have the diametrically opposite effect to that of refined cereals.

There is a view held by some that meat, and particularly the high protein content of meat, was somehow responsible for the enlargement of our brains. The assumed 'higher quality' meat diet theoretically allowed more energy to fuel the brain with a shorter small intestine. This reasoning is flawed on several fronts. Firstly, meat is supposed to be easy to

47

digest and to be a high-energy food but fruit is much more easily digested and provides more readily available energy too. Secondly, if there were a sufficient external pressure to bring about such a change as a shortening of the gut, we would expect other adaptations and changes towards a carnivorous diet as well. Certainly we would not expect adaptations to be heading in the opposite direction. Our teeth, for instance, are nothing like the teeth of a carnivore. The teeth of our nearest relative, the bonobo, are much better adapted to eating meat than human teeth are, and bonobos hardly eat any meat. In fact it is known that bonobos are, if anything, more intelligent than chimpanzees and it is chimps that eat at least some meat. So, if bringing meat into the diet of an ancestral human was enough to shorten the gut and expand the brain (both major changes), where are the parallel changes in areas that would be needed to cope with a meat diet?

If we look at areas such as dentition, the physiology to digest meat and the ability to catch it, we find nothing that looks even vaguely carnivorous. If we lined up the three most evolved species of primates – chimps, bonobos and humans – we would have to conclude that humans are in fact the least adapted to eat meat. Humans have much smaller teeth and they cannot chase the meat nearly so well. Also there is a structural distinction between carnivore guts and frugivore/vegetarian ones. Our guts are like the non-carnivores – they are folded, smooth and still significantly longer than a carnivore gut. There is a difference in saliva too. Carnivore saliva is acid but the saliva of humans is alkaline which provides the right functional environment for digestive enzymes, such as amylase, to break down starch.

Now, if we ask what sort of food really fits these human adaptations, we have to conclude it is fruit. Fruit fits the brain/gut energy equation; the shorter gut, the ease of digestion, the low toxicity and the small teeth. Fruit is easy to assimilate and the nutrition it provides is in a form that needs very little conversion to the real requirement of the brain – glucose. (The sugar in wild fruit tends to be rich in glucose and fructose compared to cultivated fruit that has been bred for its sweeter tasting sucrose content.) Humans thus have a proportionately shorter small intestine than chimps and bonobos, not because of increasing levels of meat in our diet but because of an increased specialisation on sugar-rich fruit. High quality fruit is low in toxicity and provides all the fuel the brain needs. Meat conversely is more difficult to digest, particularly without cooking, and then to turn protein into sugar requires yet more energy. So meat as an energy food doesn't make as much sense as fruit that is full of fruit sugars which are easily assimilated and take little conversion.

The anatomy and physiology of our digestive system support the case for the biochemical role of tropical fruit in human development. However, the case could be stronger still if we could show that the human brain in archaic times actually worked the digestive system in a way that extracted the nutritive elements within the plant based diets more efficiently. More research needs to be done in this area but preliminary indications (private research) hint that a digestive system run without interference from the left hemisphere may do just that.

PROTEIN, FATTY ACIDS AND WATER ~ψ~

We need to look at the whole matter of protein requirement in a little more depth. Perhaps we do not need as much as is widely assumed. The time of our life when we need the richest and highest quality of nutrients is in our first few years of life, when our bodies and brain tissues are growing most rapidly. It is surprising then to discover that human breast milk has a protein content of less than 10%. Breast milk is sweet and rich in fat, providing sugar to physiologically fuel the baby and fat to build it. It is a low protein food.

Recent research has illuminated the vital necessity of adequate polyunsaturated fats for brain development, particularly in forming nerve fibre membranes. But in their first four months, babies do not produce the enzymes needed to make certain long-chain fatty acids. The only source of these acids is in the milk they consume. The food mothers eat during their breast-feeding stage has been found to affect the balance of fats in their infants. In one study, the baby of a mother who ate a diet that excluded all animal products had twice as much polyunsaturated fat in its adipose tissue than did babies whose mothers were omnivorous. The conclusion was that babies breast-fed by mothers who ate an exclusively plant-based diet have better brain development because of the role of polyunsaturates in the growth of neural membranes. This study again points to the suitability of a fruit-based diet and its link to neural development.

In the first year of life, no less than 60 percent of a baby's energy intake fuels brain growth. Referring back to Katherine Milton's spider monkey study, we could ask whether they were really eating leaves for their protein content. They may have been primarily after additional essential polyunsaturated fats.

Fatty acids play an essential role in the structure and function of the brain. (Two of them alone, arachidonic acid and docosahexanoic acid, constitute 20% of the dry weight of the brain.) These are biologically highly active compounds that perform numerous regulatory functions in the brain and the rest of the body. Many of them can be synthesised by the body if the diet provides enough of the raw materials for construction, but some, such as linoleic (omega 6) and alpha-linolenic (omega 3) acid, are only available from the food we eat.

Wild foods routinely eaten by monkeys contain notable amounts of alpha linolenic and linoleic acid. The diet of our human ancestors would have been similarly rich in these essential fatty acids. In fact analysis of wild plant foods eaten by free-ranging primates shows that these foods are generally high in the nutrients we know are necessary for human health too. Natural primate diets contain a greater proportion of many minerals, vitamins, dietary fibre as well as the essential fatty acids than that of modern humans. It is likely then that the present recommended daily requirements for these dietary components are set far too low.

Animal studies have also shown that neural integrity and function can be permanently disrupted by deficits of fatty acids during foetal and neonatal development. These nutrients are extremely important. Research has indicated that infants may benefit markedly from the long chain polyunsaturated fatty acids naturally present in breast milk. It is highly likely that most of us are chronically short of these nutrients as they are in short supply in our modern diet and, even more crucially, are absent from many baby formula foods.

Considerable evidence is now accumulating that indicates that deficiencies in essential fatty acids is a major contributory factor in a range of interrelated childhood disorders including attention deficit and hyperactivity disorder, dyslexia, asthma, allergies and even autism. It has also been shown that correcting these deficiencies can significantly improve health.

Appleton Central Alternative Charter High is a school in Winsconsin that caters for students with behavioural and learning difficulties. In 2003 they instigated a well-being and health food program. The junk food vending machines were removed and proper lunches were offered that included raw vegetables, fresh fruit and whole grain breads. ACA staff assert that students' disruptive behaviour and health complaints diminished substantially. Students also seemed more able to concentrate. They became more stable too, so those mental health and anger management issues were easier to manage. Teacher Mary Bruyette said she saw changes 'overnight'. She noticed a considerable decrease in impulsive behaviours, such as talking out, fidgeting and the use of foul language. Henceforth she has had fewer disciplinary referrals to the office for students who could not settle down and do their coursework. Complaints of headaches, stomach aches, and feeling tired also lessened. Students were no longer hungry mid-morning or mid-afternoon. According to Principal LuAnn Coenen, negative behaviours such as vandalism, drug and weapons violations, dropout and expulsion rates, and suicide attempts are now virtually non-existent.

The school also experimented with junk food days – days when the students reverted to a diet of chips, brownies, candy bars and sugared sodas. Students became tense and 'wired'. They were unable to focus and complained of stomach aches and tiredness. Students and staff mutually agreed to abandon such days. The negative effects of such junk food have been further highlighted in the recent shock film – 'Super Size Me'.

Just replacing sodas with water can make a significant difference. Most humans today are chronically dehydrated. This simple fact causes much ill health. According to Dr F. Batmanghelidj (in his book 'Your Body's Many Cries For Water') many of the degenerative diseases of the human body are caused by a simple lack of water. He has concluded from his studies that asthma, diabetes, arthritis, angina, obesity, Alzheimer's, high cholesterol, hypertension, dyspeptic pain and many other maladies are signals from a body that is desperately thirsty. We are much more prone to dehydration if the bulk of the food we live on is dry. Fruit and vegetables have a much higher water content than grain and wheat-based products. That our bodies work more efficiently when we live on the diet that provides not only the nutrients that we need, but also basics like our water too, is further evidence for this being the one we have been 'designed' for.

SUMMARY ~ψ~

In this chapter, we have presented the framework for an alternative mechanism that we think was fundamental to the evolution of humans, hominids and perhaps to the great apes too. We have argued that humans are best adapted to a diet high in fruit and that this diet played a significant role in our development. Primates evolved in an environment with a unique biochemistry. This environment may have been stable for millions of years, and, over such long stretches of time, this rich chemical matrix would have had a very real effect. If in the distant past, humans and proto-humans ate a diet consisting mainly of fruit, then the chemicals contained within the fruit would have flowed through their bodies for countless generations. This biochemical influence could have caused, for instance, a lengthening of the juvenile period and much else besides. In the next chapter we will look in detail at these biochemical pathways and how they acted on the human system.

CHAPTER THREE

~ψ~

Figs, Steroids and Feedback Loops

A revised theory of inheritance explains many human mysteries - why are we so different from our closest animal relatives, how our brains became so large but then started to shrink and why the two sides of our brains apparently have different functions. We find that, far from a continued advance, the human system has suffered a stall in its development and this has affected our health, how we feel and even how we behave. Identifying this problem is the first stage of finding a solution.

THE CORE HYPOTHESIS ~ψ~

There is no satisfactory explanation for why humans are physically, mentally and culturally so very different to bonobos, chimpanzees and gorillas, when, as far as our genetic blue print is concerned, we are all nearly identical. Chimpanzees are genetically closer to us than they are even to gorillas, despite their closer physical resemblance. The uniqueness of humans therefore requires a serious explanation that does not strain our credulity by invoking 'outerworld' concepts that can never be proved or tested. We suggest that the unique features of human evolution and our expanded consciousness can be best explained by a mechanism that acted slowly over the millions of years of primate evolution but, somewhere down the ape/hominid lineage, started to 'sky rocket'. This mechanism was driven by the biochemistry of a predominantly plant-based diet that acted on the steroid hormone environment of the animal in a way that altered steroid activity.

So what are steroids and why are they so important? Steroids are fat-soluble organic compounds that occur naturally throughout the plant and animal kingdoms. They include molecules like cholesterol, which in animals are transformed by a series of biochemical steps into specific hormones. For example, enzymes in the male and female reproductive organs

53

change cholesterol into the familiar 'sex' hormones – testosterone, progesterone and the oestrogens. Other enzymes convert cholesterol into other kinds of steroid hormone, such as cortisol, which is secreted by the outer layer of the adrenal glands in response to stress. These hormonal steroids pass through cell walls and act deep within, in the nucleus, where they regulate the 'transcription' of various genes. Transcription followed by the translation process results in the construction of proteins, and these are the building blocks of our cells (structural proteins) and the chemicals that run them (enzymes).

Hormones alter cellular operations by changing the types, activities or quantities of important enzymes and structural proteins. They can stimulate the synthesis of proteins and enzymes; they can affect the activity of enzymes by turning them 'on' or 'off' and they can increase the rate at which the various proteins and enzymes are made. By these mechanisms, hormones can modify the physical structure and biochemical properties of the cellular system. Hormones are a fundamental part of our functioning – our very structure depends on them.

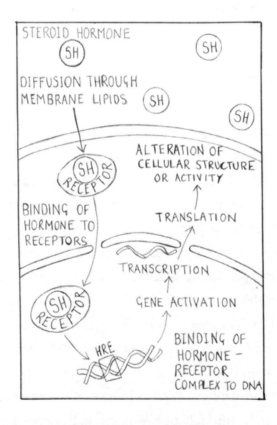

Fig 4a: This diagram illustrates the basic transcription process and highlights the role that steroids play.

Steroid hormones in particular are an integral part of the mechanism that reads the DNA, the blueprint that ultimately dictates the structure and chemistry of what is built and ultimately how it works. To make this clear we can consider the different developmental pathways that lead to a male or a female.

If a developing individual has a Y-chromosome, it will become a male; otherwise it defaults to the female developmental pathway. The Y-chromosome's primary purpose is to initiate the production of a protein called 'testes determining factor' which induces undifferentiated foetal gonadal cells to develop into testes. Once this occurs, sexual dimorphism is not driven by the presence or absence of the Y-chromosome but by the chemicals secreted by the testes. These include a peptide hormone which inhibits the development of a female reproductive tract and the steroid hormone testosterone (together with the closely related estradiol) which effects changes in the neural structure of the foetus and directs the masculinisation of the entire body.

Once the DNA has set the developmental pathway in motion, the Y-chromosome no longer has a significant role. It is the hormones that actually effect the changes. Their importance can not be over emphasised.

In transcription, the key players are the code (DNA), the reading equipment (ribosomes) and steroid hormones which, in simple language, tell the ribosomes what to read. At puberty, the steroid hormones tell the ribosomes to read the code differently from before and what results (in males) is a new structure with bigger chests and more body hair. It's the amount or activity of the steroid hormones that dictates what happens. Steroids are extremely powerful chemicals. Minute changes in the levels of steroids in the body can have a big effect. Any external or internal factor that changes the balance of these chemicals can have major consequences. We can see from individuals wishing to change sex that, even after the body is built and mature, taking more or less of a particular hormone can result in structural changes. Establishing a different hormonal regime can for example result in men growing breasts. The DNA code has remained the same but the structure of the body has changed.

There is a major degree of plasticity within this building mechanism, and this is at its greatest in the earliest stages of human growth and development. The effects of an abnormal hormonal environment during pregnancy have been well documented. For example, in cases in which the foetus is a girl and the mother, for whatever reason, has expressed abnormally high male hormone levels, babies have been known to develop pseudo male genitalia. These aberrations can be treated by surgery and hormone therapy but such androgenised girls also display a whole host of masculine behaviours as they are growing up. They have been found to expend significantly more energy during play than other girls, prefer boys as playmates, engage in more fighting, and have much less interest in dolls, play motherhood, make-up and clothes. This plainly demonstrates not only how structure can be altered by a change in the balance of an individual's steroid hormones, but also how an altered balance of a mother's hormones can affect her baby permanently.

Since the elucidation of the DNA molecule, a great deal of attention has been focused on its central role as a blueprint for growth, development and functioning, as well as the evolution of higher organisms. But, on its own, DNA is of no more use than is the hard disc of a computer. If anything, it is the equipment that reads the disc that is more complex and central to the end result. We need to remind ourselves that in considering the evolution of the biological system, the interaction between DNA and the transcription system is a more useful framework than just DNA on its own. Once such a framework is established as possible, a significant variation on the standard model becomes evident.

DNA is usually thought to be the only conveyance for the passage of information to the next generation and hence to variation, adaptation and evolution. The mechanism that reads the DNA is assumed to be stable. This is not usually included in the picture as far as inherited change and variation is concerned. The standard evolutionary model is based on the changes that come from glitches in the DNA code. These changes are taken to be accidental (mutation) and usually deleterious, but when they are of benefit, the benefits incurred will create a fitter animal with enhanced survivability (selection). However, there could be, theoretically at least, a secondary mechanism for inheritance. If variation in 'what is built' within an animal's lifetime somehow affects how the DNA is read, then these slight differences may be built into the structure of an offspring in the next generation.

We propose that something acting on the functioning of this mechanism gave rise to the unique features of primate evolution, eventually reaching its maximum expression in humans. Without contradicting accepted biology, a case is presented for externally driven steroid and monoamine oxidase (MAO) inhibition as the significant missing piece of a puzzling jigsaw. (MAO is a key enzyme that regulates neurotransmitter activity by breaking down key neurotransmitters.) The importance of this has not been fully explored before because of the continued emphasis on DNA as the only variable part of the evolutionary inheritance mechanism.

Loops within loops

A fundamental property of our, and all animals', biochemical systems is their extreme delicacy – only a very small amount of chemical change (as we can see from the contraceptive pill) can have a profound affect on function. If the diet of our primate ancestors in the tropical forests was full of biochemically active material some effect was inevitable. As many of the chemicals in fruit are known to modify steroid activity, it is very reasonable to suppose that, over millions of years, they were responsible, or at least partially responsible, for many changes to functions like the timing of sexual maturation.

The way hormones, steroids and the DNA reading mechanism function are complex, for they interact within a web of interdependent loops. The neural network and ultimately the brain are significantly involved too. The glands, which produce the hormones, were once thought to run independently and the levels of production self-regulating – the hormones themselves regulated hormone levels: but now specific neural pathways have been found to end in the glands. These are believed to directly stimulate the production of hormones – the glands are thus not autonomous. The neural system runs or at least modulates the hormone system. There is a tendency within orthodox science to the belief that discrete systems work in isolation. This is due in part to the way subjects are studied in isolation. (The study of the ductless glands and their secretions is termed endocrinology.) But now there is an absolute acceptance of the link between the neural and endocrine systems, so much so this branch of investigation is now labelled 'neuroendocrinology'. We now know that the brain affects the hormone system that in turn regulates the DNA reading which in turn influences the animals' structure and the mechanism of construction.

This is important, for slight changes in the construction of the brain could affect the modulation of the hormone system. An altered brain, for example, could result in an increase or decrease in the quantity of hormones, like melatonin, that the pineal gland produces.

A progressive hormonal effect

Though these interconnecting factors are far from simple, so far we have shown the importance of steroids as regulators of DNA transcription, and how the level of steroid sex hormones can affect major changes in the body. We have noted the sensitivity of the body's systems to these powerful chemicals and pointed out that the brain is connected to the hormone producing mechanism too. We have also considered how an external factor, such as diet, can affect the levels, or at least the activity, of these substances in the body. We will now show how these factors could have played a part in the evolution of our big brains, via a hormonal effect initiated in one generation creating more of the same effect in the next.

In the uterus, the hormonal environment provided by the mother affects how the foetus is built by acting on its DNA reading mechanism. The neural and endocrine systems of the foetus will be changed by this mechanism. Furthermore, a variation in the building programme will not only affect the function of these systems in that individual's lifetime, but could also affect the next generation as well, because, if the offspring is a girl, it could alter her own future uterine environment.

The changes brought about by an altered hormonal environment in the uterus will be concrete and structural. What is constructed in the baby in say weeks 3, 7 and 9 of pregnancy will be lasting. This offspring will have a different structure and function from its parents due to the different hormonal/chemical balance it was exposed to during gestation. The big variable here is the reading mechanism not the DNA code. We have seen in examples, such as the physical changes that occur at puberty, how central the reading mechanism is to structure. It is not the code that is changing at these times, it is how the DNA code is read, and this is chemically influenced. The way steroids act is the variable link in this mechanism.

This is a crucial piece of our hypothesis, so in the interests of clarity, we will run through this once more in a slightly different way.

If, for example, a pregnant woman had a slightly abnormal hormone regime, it would be this regime which coursed through the body of her foetus. The growth and development of the foetus would be slightly changed in response to this altered regime. Even though its DNA code is unique, the growing child is not an autonomous unit growing in isolation. In the uterus, she (and it is the female line that is significant) is being flooded with her mother's hormones, and this will have some effect on aspects of how she is built.

There could be a number of slight changes brought about by this hormonal effect but for our theory we need to focus on the possibility that it is the foetus's neuroendocrine system and hence its DNA reading mechanism that is modified. These structural modifications would then be with that individual for its entire lifetime. This is of particular importance, for the neuroendocrine system will affect the growth and development of the child, including potentially the length of the juvenile period, the timing of puberty and how her brain functions. It will also affect the hormonal environment in her uterus when she conceives. The hormonal regime that floods her uterus will be different from that of her mother because she has a neuroendocrine system which is structurally different. She will pass her normal, unchanged (by this mechanism) DNA to her offspring but, from day one in the uterus, the new foetus will develop in a slightly more altered hormone environment which will again affect how its neuroendocrine system and DNA reading mechanisms are built.

The DNA is not being altered by this mechanism. But there is an effective DNA change because what is built is dependent on how the DNA is interpreted. If these changes were all generally flowing in the same direction, what would be built could change progressively over the generations.

There could have been thousands of different variations within this overall framework that went nowhere, but if in one lineage there was an inhibition of steroid activity that in turn led to more steroid inhibition through subsequent generations, this one variation could have led to lasting change. If such structural changes resulted in the production of, for example, more melatonin and a modified steroid environment, the loop could start running faster. More melatonin and less steroid activity boosts this process because melatonin suppresses steroid activity, and reduced steroid activity takes the brakes off melatonin production.

Such changes could lengthen the juvenile period. And, because neural development, in effect, stops at puberty, a longer juvenile period allows more time for the development of the brain and indeed the neuroendocrine system. It is theoretically possible then that this longer window of development boosted pineal activity and hence melatonin production, affected the neuroendocrine system, and had a knock on effect on the uterine environment eventually provided for that person's developing child.

These are key factors. The mechanism will only work if these related factors flow in the same direction – in the direction of suppression of steroid activity. It is also necessary for there to be an incremental increase with every generation. This may perhaps seem unlikely, but we are trying to explain an unlikely developmental event (the production of a uniquely big-brained primate). All we need for a fast track mechanism to get up and running is the DNA reading system to change in a way that produces ever less steroid activity.

A NEW THEORY OF INHERITANCE ~ψ~

What we are proposing here is nothing less than a new theory of inherited and evolutionary traits. The standard model is totally DNA based: inherited traits are passed on via DNA codes but, if a different reading system can be inherited and passed on, there is, in effect, a transmission of a different DNA expression. The DNA and the reading system do not work in isolation. They go together. The reading system is built in the uterus. A change in this reading system will result in different structures, including the structures that read the DNA. If these are stable, the way the DNA is read will be changed permanently. This is a new and radical theory that has huge implications. It is a mechanism for inheritance that does not depend on changes in DNA. It is an inherited reading change.

This theory is not incompatible with the standard DNA model for inheritance. It is merely a variation that, we propose, had a marked affect on the evolution of the ape/hominid lineage. The key point is that the variation is coming from the neural-endocrine system and it is this variation which is inherited. However this does not preclude DNA variation working with, alongside, in response to, or independently of this mechanism.

This mechanism is really very straightforward and it is surprising it hasn't been identified before. At its simplest, we can see that, as we grow, our DNA/transcription system builds the brain. The brain regulates the hormone system, and the hormone system, particularly the steroid hormone system, is part of the transcription of DNA mechanism. This forms a circular loop. Anything entering this loop, for instance a biochemical influence from a fruit diet, will affect all parts of it in the following ways:

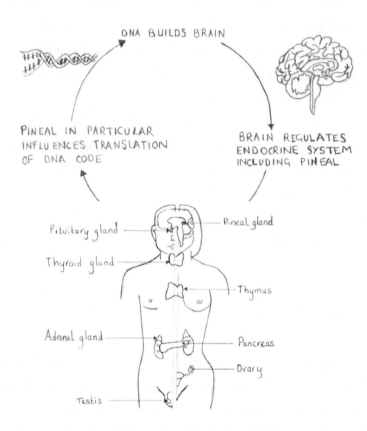

Fig 4b: This diagram shows the major players in the transcription mechanism. The DNA code builds the brain, the brain regulates the hormone (endocrine) system and the endocrine system regulates steroid activity. In most higher organisms, the neuroendocrine system is a fairly stable part of the mechanism; changes that occur originate in the DNA code. We will demonstrate a means by which the neuroendocrine influence on transcription can change and pass from one generation to the next without changes in the DNA code.

(a) Steroid inhibition directly affects transcription. An individual growing and developing in an altered transcription environment would be built slightly differently. This difference would include an altered brain. As we have seen, the brain modulates the hormone system. If the altered brain affected this modulation in a way that suppressed steroid activity, a loop of progressive effects could be established.

60

(b) The individual's juvenile period would be extended due to the direct effects of the sustained steroid suppressing chemicals. An extended juvenile period would allow for a longer period of brain growth (steroid hormone activity at puberty brings to an end the neural development). This could again lead to a slightly different brain structure and function.

(c) The direct effects of a fruit diet, rich in MAO inhibitors, are likely to stimulate greater pineal activity. The pineal would produce more melatonin and beta-carbolines. These chemicals have a similar effect to the chemicals found in fruit. They suppress steroid activity.

Any neural changes created by this mechanism would initially have been very slight, indeed minuscule, in any one generation. If however a build up of change caused the pineal to produce more chemicals which reinforced the external chemical effect then the rate of change could increase. We propose that this loop of change built to a point in which the pineal became the dominant effect. So, while this process would still have been underpinned by a fruit diet that provided continuing supplies of steroid suppressing chemicals, the predominant effect became increased pineal activity. This significant biochemical change could have created an internal environment that, via the stages we have elucidated, produced an even larger brain with enhanced functions, and the potential for a slightly bigger brain with each successive generation.

Over the last few million of years, there was a slow increase in neural capacity in primates, largely we believe as a result of increasing pineal activity initiated by a diet rich in fruit. In the human line the pace of expansion quickened. There was a doubling of brain size in a very short time. The biochemistry of plants has a lot to answer for.

CHEMICALS IN PLANTS ~ψ~

Plants in general, and their fruits in particular, contain a very large number of chemicals. They could with great justification be called biochemical factories. Many of these chemicals are similar to ones within the human body and can affect how our internal biochemistry works. There is a close correspondence between the plant and animal biochemistry, which is why we utilise plants for healing and as the basis for many pharmaceutical medicines.

Two new breast cancer drugs, (tamoxifen and exemestane) are close to synthetic equivalents of plant chemicals called flavonoids. Exemestane works by blocking the action of the enzyme aromatase, which converts androgens into oestrogens. Tamoxifen directly inhibits the activity of oestrogen. As most breast cancers cells need oestrogens to divide and grow,

61

these actions can stop cancer growth and even cause tumours to shrink. In the development of the drugs it was found that some naturally occurring flavonoids could powerfully inhibit the action of aromatase too.

Another study further emphasises the power of flavonoids. Dr Richard Sharpe, Senior Scientist at the Human Reproductive Sciences Unit in Edinburgh, was concerned about the potential effects on babies of endocrine disrupting flavonoids in Soy milk formula. Marmosets were used in the trials as they have a similar endocrine system to humans. The study showed that simply feeding a flavonoid rich diet post-nataly was enough to severely inhibit the neo-natal testosterone surge. (In humans this surge has been linked to neural development and synaptic pruning.) Though this study is ongoing, initial findings suggest that the marmosets are not adversely affected by the flavonoids. From our perspective however, the main question that arises is – if these chemicals were once present in our diet all year round for millions of developmental years, what affect has their loss had upon us? Perhaps this testosterone surge would have been ameliorated.

Another group of plant biochemicals, the beta-carbolines, not only inhibits the action of steroids but also elevates neurotransmitter activity. Beta-carbolines, as well as occurring in some fruit and leaves, are also endogenously produced in the pineal gland of animals. If taken orally, large doses of these chemicals can even be hallucinogenic.

Within the body, beta-carbolines act as neuromodulators, fine-tuning the actions of neurotransmitters. They do this by preventing the chemical monoamine oxidase breaking down the neurotransmitters serotonin and noradrenaline. Beta-carboline action can thus result in a build up of these neurotransmitters at the synapses (the junctions of nerves) which allows greater neural activity. Indeed this is just how hallucinogens work.

There is one specific beta-carboline (6MeOTHBC) that shares some of the properties of melatonin. Like melatonin, it shows a circadian rhythm within the body, it inhibits the development of the genital organs and in females can even interrupt the oestrus cycle. It is also produced in the pineal gland and has now been given the much more accessible name of 'pinoline'. Pinoline is a very effective antioxidant, particularly good at free radical scavenging in brain tissue, and it can also act as a very safe anti-depressant. It does not have the side effects associated with Prozac and other serotonin-based drugs and works by increasing serotonin turnover and recycling other essential neurotransmitters.

The harmala alkaloids are an extremely interesting group of beta-carbolines. Chemically they are very closely related to pinoline. These are found in the South American vine, *Banisteriopsis caapi*, which is traditionally used in conjunction with a number of plants containing DMT by tribal shaman to induce visions (more on DMT in Chapter Five). The specific alkaloid, which is the active ingredient in the *Banisteriopsis* brew known as

Ayahuasca, has even been given the name 'telepathine' on account of its profound properties. The same alkaloid also occurs in the Assyrian Rue, *Pegunam harmala*, which is a plant with a very similar ancient healing/visionary pedigree. Passion Flower, *Passiflora incarnata*, a common ingredient in herb teas, contains these chemicals too but in weaker concentrations.

Fruits contain chemicals that are neurotransmitter precursors. These are chemicals which animals can utilise intact or which need little conversion to build neurotransmitters. They include plant hormones like indole-acetic acid and tryptophan. Tryptophan is a key amino acid that is found in protein rich foods like meat but it also occurs in some fruits like banana and avocado. In Polynesia, fruits rich in tryptophan, like Noni, have been traditionally used in healing and tryptophan has even been used as the key ingredient in one type of sleeping pill. This amino acid is needed by the brain to make serotonin and recent research has shown serotonin to be a body chemical of great importance. Its range of functions include easing pain and tension and acting as a relaxant.

Fruit (as well as vegetables, nuts, seeds and flowers) contains a whole host of other chemicals which, while not being so powerful as beta-carbolines, will have some affect on the biochemistry of the consumers. Recent research has found that these substances have antibacterial, antifungal, antioxidant, antinflammatory, antimutagenic, and antiallergenic properties and they can inhibit the activity of several enzymes too. In experiments on mice, it has been found that many flavonoids inhibit monoamine oxidase (MAO) and the potency of one, apigenin, was found to be comparable to that of clinically used MAO inhibitors (used to remedy human depression). Another flavonoid, quercetin, which occurs in nearly all plant foods and gives the colour to, for example, apple skin, appears to have properties that reduce the risk of coronary heart disease and prevent ulcers, cataracts, allergies and inflammation.

Dr Richard Wurtman, a neuroendocrinologist at the Massachusetts Institute of Technology, spent years investigating neurotransmitters. After tracking the pathways they followed and studying their behaviour and interactions, he concluded that the brain's ability to make certain neurotransmitters depends on the amount of nutrients circulating in the blood, and this is intimately influenced by what we eat. Fruit, therefore, as a rich source of these chemicals, can have a profound affect on our neural systems.

It is also of consequence that fruit is a rich source of simple sugars that do not need much alteration to make them useful to animal metabolism. They are easily converted to glucose – the number one brain fuel. This may be a factor of great importance. Fruit not only provides the chemicals to change brain activity but also the fuel to run it.

This short review of just a small fraction of plant chemicals serves to highlight their potency. Over an evolutionary time scale a perpetual drip of these fruit chemicals via diet would have affected any specialist fruit feeder. They represent a potentially powerful force for change. Flavonoids in particular are extremely potent endocrine modulators and they would have been, in effect, an integral part of our ancestors' endocrine system for millions of years. Flavonoids powerfully inhibit both the activity of steroids and the conversion of androgens to oestrogens. They also inhibit the action of the enzyme monoamine oxidase (which could have taken the brakes off the body's production of melatonin.) The fruit of the forest, via this linked complex of biochemical action, may have been just enough to initiate primate development and push this process forwards towards the genesis of our own species.

THE FIRST PRIMATES ~ψ~

The primate story began during the Cretaceous period, some 70 million years ago, when small insect-eating mammals, that may have resembled the present day tree shrews, climbed into a new rich foliage to forage amongst the flowers, leaves and fruit. Flowering trees were spreading across the globe and the emergence of these plants brought a new level of complexity into the evolutionary equation. The more primitive plants, which came before, were not only less chemically rich but also disseminated their seeds without producing much in the way of fleshy fruits.

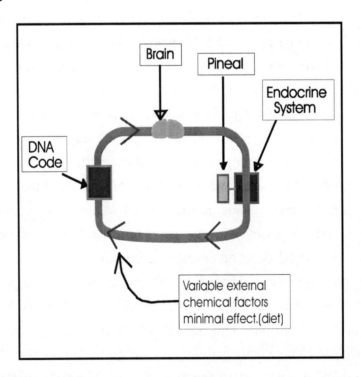

Fig 4c: This shows the basic components of the inheritance mechanism. The external dietary factors have minimal or zero effect on inheritance, as is the case in virtually all animals.

Like all living creatures, the DNA of this tree climbing mammal would have dictated all aspects of its structure from molecular, sub-cellular, cellular and neural levels through to its gross anatomy and physiology. Similarly, its steroid hormones would have dictated the reading of the DNA codes. This is the standard biological model of life. Evolution runs on DNA variation, inheritance and selection. But here is the key; anything entering this relatively stable loop could have an effect.

A CHANGE OF DIET AND AN INCREASE IN BRAIN SIZE ~ψ~

At some stage the diet of the insect-eating arboreal primates changed to flowers, leaves, fruits and nuts. This massively changed the biochemical base of the species. Something that had been constructed of insects now constructed itself from plant material. This diet enhanced brain growth by suppressing steroid activity, boosting neurotransmitter activity and providing sugar rich fuel. Thus from tree shrews to apes, over millions of years, there was a slow but continual increase in brain size and function. The diet also had the tendency to extend the juvenile phase and this allowed a longer period of brain growth before the sexual hormones kicked in and ended this phase.

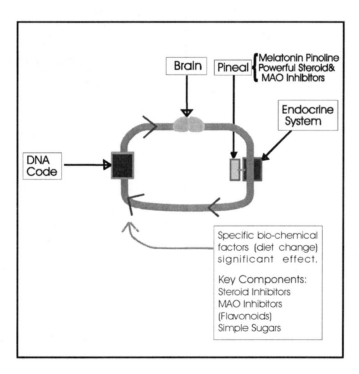

Fig 4d: Here we see a change in diet that brings a significant hormonal effect into the loop. There is a sustained inhibition of steroid hormones.

So there were two linked forces of change: an increase in the juvenile period, which allowed for an increased period of brain growth, and the direct effect of the chemicals present in the fruit on internal biochemistry. Steroid activity in particular would have been reduced.

The external influence of the forest chemicals was primarily responsible for the unique neural evolution of primates. An increasing synergy of powerful hormonal effects entered the standard mechanism of inherited change and provided a means of rapid evolution without the need for atypical genetic variation. The key components provided by the flowering tree 'factories' included steroid and MAO inhibitors, neurotransmitter precursors and simple sugars. The combination of these active ingredients working on the hormonal system of the primate resulted in increased neural activity and an expanding brain. Because of the way the hormonal environment of the uterus was altered, these changes could be passed to the next generation without a change in the DNA code.

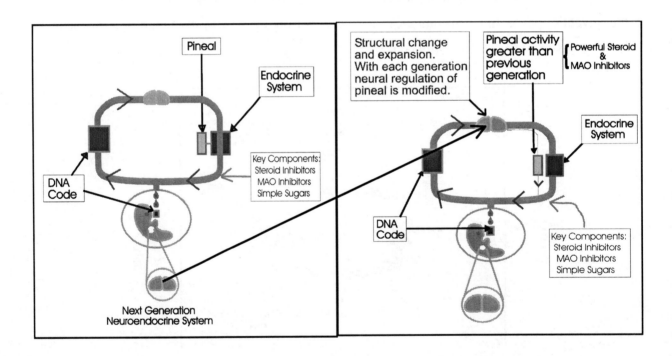

Figs 4e and 4f: Each generation of babies are exposed to greater steroid inhibition than the previous generation. Each adult generation then produces greater steroid inhibition than the last. Each increase produces further increase and this combines with the static dietary effects to create a positive feedback mechanism.

Neural development in the uterus is particularly sensitive to the hormonal environment. We saw in our first chapter how testosterone in particular affects brain development and how small changes in the level of this hormone can lead to autism. The hormonal environment in the uterus therefore not only affects the structure of the developing neural system but also its future functioning. Over evolutionary time, more positive structural and functional changes occurred. The hormonal environment in the uterus of a female primate/hominid subject to these changes would have been a notch up from her mother's, and thus there would have been an increased affect on her own offspring. This mechanism led to larger brains and more active pineal glands.

THE PINEAL AND EXPONENTIAL BRAIN EXPANSION ~ψ~

With a continued fruit diet, the sustained external effect that had gone on for millions of years finally brought the neural system to a point of fundamental change. A substantial increase in the output of the pineal gland was the pivotal factor that led to the increasingly rapid expansion of the human brain.

The pineal produces a range of very powerful chemicals that inhibit steroids and elevate neurotransmitter activity. It contains special secretory cells called pinealocytes that synthesise the hormone melatonin from the molecules of the neurotransmitter serotonin (sometimes called the 'happy hormone'). Melatonin is known to slow the development of sperm, eggs and reproductive organs and it plays a major role in the timing of human sexual maturation. Melatonin levels in the blood decline at puberty and if anything damages the gland so that melatonin production is eliminated, puberty will be initiated prematurely. Melatonin has many other properties too. It is a very effective antioxidant protecting nerve cells from free radicals, it regulates sleep patterns, strengthens the immune system and can protect the body from cancer, heart disease and stress. Recent work by Professor Igor M Kvetnoy in Russia has shown just how ubiquitous melatonin is within our bodies and what a wide range of effects it has. Among these are control of biological rhythm, influence of the reproductive cycle, immune response, monitoring of mood and sleep, and cell differentiation and proliferation.

At some point the hominid brain, primed by an evolutionary history of ingesting the complex of chemicals in fruit, reached something like a critical mass of activity. After this fulcrum was passed the pineal pumped increased amounts of chemicals like melatonin and pinoline. These chemicals (and this is the clever bit) are the very ones that further accelerate the whole process. Thus brain activity was elevated to a higher degree still. The chemicals the pineal produces are similar to the chemicals found in forest fruit but they are more powerful and available in larger amounts. More of these chemicals will tend to further extend the juvenile period too so that the brain can grow still larger. If this in turn elevates neural activity, and if the elevated activity stimulates the pineal gland into greater production, the whole loop will be cranked up yet another notch. So once the process is initiated, the bigger brain and greater neural activity boosts the pineal, which results in an even bigger brain and more activity. The pineal is stimulated to work harder resulting in a rapid increase in brain size, activity and function. The more powerful the pineal becomes, the more chemicals it pumps causing more size, more activity, and still more chemicals – a wonderful feedback loop.

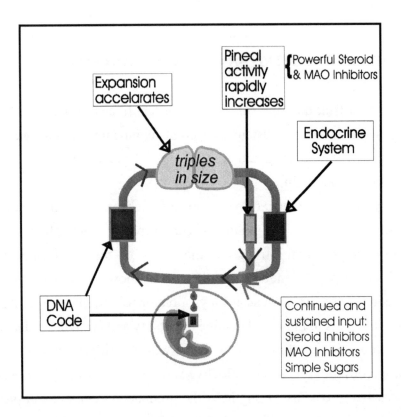

Fig 4g shows the mechanism in full flow. The increase in pineal activity has become the most powerful steroid inhibiting factor but the dietary factor is still relevant.

A unique process thus had a unique result – the production of a huge-brained primate, the like of which had never been seen before on earth – *Homo sapiens*. And, of course, this brain was not just large but of a different type. The whole exponential activity created a new kind of brain that was increasingly dominated by a disproportionately large neocortex.

In normal evolutionary terms, the doubling or tripling in size of the brain over such a short time scale is absurd. It is highly unlikely that the DNA selection process could have achieved this rapid result (certainly not on the savannah where hominids would be subjected to hard and stressful environmental conditions). Some unique process must have been at work, and this hypothesis fits the bill.

SURVIVING ON THE SAVANNAH ~ψ~

Some time around 200,000 years ago, something dramatic occurred. The already shrinking forest may have been subjected to some further catastrophe (meteoric, volcanic or both) which forced the last of the forest-dwellings humans out. Other Homo species, sub-species and lineages may have left the forest earlier but the lineage which gave rise to *Homo sapiens sapiens* remained in their benign home until around this date (give or take 50,000 years or so). It had been an environment that stimulated and supported maximum neural development but now our ancestors had to leave to find food elsewhere.

The change would have called for a rapid initiation of survival strategies. The migrants would have been subjected to greater extremes of temperature on the harsh savannah for the days would be hotter and the nights cooler than back in the womb-like forest, and the ultra-violet component of the sunlight would cause skin damage too. They would almost certainly have been subjected to competition from other hominids living there already that were well adapted to life as hunter/gatherers on the plains. And they would have to find different sources of nutrition. The food available – insects, shoots, tubers and meat – would lack the biochemistry which propelled the rapid expansion of the brain. But, despite these major challenges, they survived, for their one advantage was a brain with the capacity to solve such problems.

An enforced change of diet would have had a major effect on the hormone/ brain expansion loop that depended on a foundation of forest fruit chemicals. Losing the biochemistry, which underpinned this cycle, for even a single generation may have been enough to send the loop into a downward spiral. Tubers and leaves may have provided some of the chemical nourishment needed but, compared to fruit, they lacked the quantitative and qualitative richness. And a diet that included insects and meat would contain a very different biochemistry. It would have provided the energy to survive but the elements, which made up the earlier unique regime, would have become much weaker. Some parts of the new diet would actually be negative. For example, meat contains the very same hormones that the chemicals in a fruit-rich diet had so successfully suppressed.

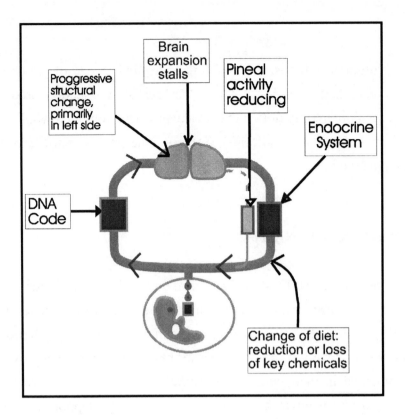

Fig 4h illustrates a change of diet and the loss of the external steroid inhibition. Though proportionally a small factor by now, its loss is still sufficient to create a net loss of steroid inhibition during foetal development. The next adult generation produces less steroid-inhibiting factor than previously. A positive feedback loop, now dominated by increasing pineal activity becomes a negative feedback loop dominated by reducing pineal activity. Loss of steroid inhibition creates further loss with each generation.

Changing to a 'survival diet' would have had a range of linked effects. A decline in fruit sugars would have reduced the fuel to run the brain, and a reduction of externally available monoamine oxidase inhibitors and neurotransmitter precursors would have again reduced brain activity. This in turn would have suppressed pineal function. If the pineal produced lower quantities of steroid-suppressing chemicals, the brain would be subjected to higher levels of functioning steroids, reducing neurotransmitter activity still further. This reduction in activity could have led to a further reduction in pineal activity because of the link between the pineal and the brain. Thus a downward spiral could have been initiated by the removal of a fruit-rich diet and its replacement with a diet which included the very steroids the fruit chemical had suppressed for aeons. In the forest, the pineal pumping these steroid-suppressing chemicals had become the most important loop in a complex of feedback mechanisms underpinned by the fruit effect. With a change of diet the foundation of these mechanisms was lost.

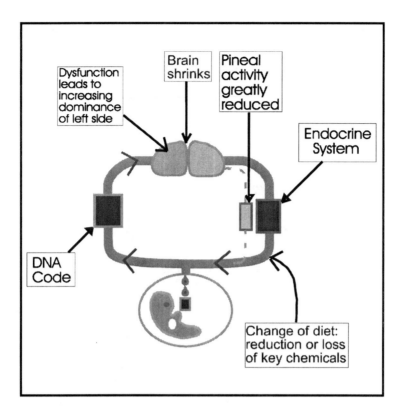

Fig 4i illustrates the negative feed back mechanism gaining momentum. The normal hormonal environment is sufficiently changed to alter neural development of the foetus. The deleterious structural changes in neural development predominantly affect the left side of the brain and cause the pineal to produce lesser amounts of melatonin and pinoline.

Reduced fruit intake leads to less brain activity which leads to less pineal pumping which leads to less function. This is exactly the reverse of that which was happening in the forest. The pineal pump was the dominant force in boosting neural growth and activity, but once this effect had begun to decline, even putting back the fruit elements would not have put the mechanism back on track. Reverting to a fruit diet, even within a generation or two, may have slowed the negative shift but it would not halt it altogether because the pineal effect had become so much more powerful than the effect of the external dietary influence. Feedback loops both positive or negative, once underway, have a lot of impetus. The fruit effect was necessary to keep the pineal pump running in one direction but once this direction was reversed it just became one weak effect in a basket of biochemical influences.

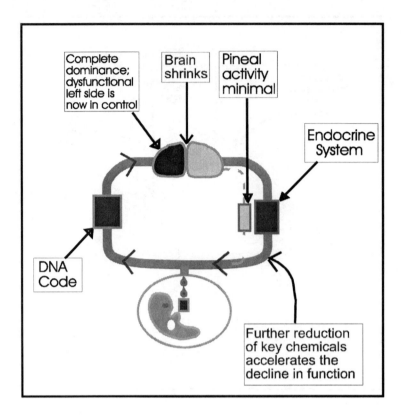

Fig 4j illustrates a significant reduction in the internal inhibition of steroids. The dysfunctional half of the brain gaining an increasingly dominant role has exacerbated this. Its dysfunction is such that it is unable to regulate and maintain a highly active pineal. Increasing dominance hastens the negative feedback loop leading to ever greater dysfunction.

Divide and rule

So dietary change reduced pineal activity resulting in less production of chemicals such as melatonin. Externally, the quality of materials used to build the neural tissue declined, and the new brain, the product of a few million years of evolution in low steroid conditions, was subjected to a rapid and progressive loss of steroid inhibition.

Slowly there was a return towards an archaic internal steroid environment as the pineal, incrementally with every generation, suffered a reduction of activity. This would have had a major knock-on effect, for the body's steroid environment fundamentally affects the way cells grow and hence how they function. Such a new internal hormone regime would have changed everything from pineal performance to perception. Even the extending juvenile period would have stalled and slowly begun to reverse, giving the brain less time to develop. With an ever less ideal biochemistry, there would have been an ever less ideal uterine 'building site' too, and an ever less ideal product. A neuroendocrine system with declining function will build one with an even further reduced function.

Something else happened during this cusp of change that was to have severe consequences. The effects of reduced melatonin and increased steroid activity were lateralised between the brain's hemispheres. An archaic genetic variation in steroid sensitivity between each hemisphere was now amplified as the neural tissue that had proliferated in a relatively low steroid environment was exposed to higher steroid activity. The steroid influence on DNA transcription started to change delicate sub-cellular structures, particularly in the left hemisphere. This had a detrimental affect on neural function and perception.

Before this time, despite some asymmetry, humans had a single consciousness system. Both hemispheres were balanced and had basically the same function. But under this detrimental regime, one half of the brain started to be affected more quickly than the other, or perhaps one half of the brain was affected exclusively. All this may have not been quite so catastrophic if the more damaged side of the brain drifted into obsolescence but quite the opposite happened. It began to take a bigger overall role whilst, at the same time, its function began to decline. The more influence it assumed, the more it fuelled the decline of the whole system. Unfortunately it is the nature of these loops that the more dysfunctional a system, the more difficult it becomes to maintain the biochemistry that would have slowed that dysfunction. If both sides of the brain had been affected equally so there was a uniform degeneration, putting the tropical fruit influence back may have made a bigger difference. But if the two halves of a system were in conflict and the half that was degenerating most quickly ended up taking the biggest role, the mechanism was set in place for severe negative changes.

At some point the lowest function took over – the left hemisphere started to run the show – dominating our sense of self and ineffectively running the neuro-endocrine, immune and other systems. The resulting biochemical regime added to the problem with each subsequent generation. The right brain was less damaged. It may still retain the potential of its original function but its influence has become limited. The left side for some reason became dominant.

Why did the altered steroid regime affect the two hemispheres differentially? Studies on cerebral asymmetries have been made on humans, fossil anthropoids (by examining the shape of the intracranial cavity), apes and monkeys. They have shown that asymmetries do exist in present day apes, monkeys and in ancient hominids, but that the differentiation is most marked in *Homo sapiens*.

This structural differentiation would seem to suggest there were functional differences between the hemispheres (as there are in present day humans): for instance, the two sides of the brain may have had asymmetrical responses to certain chemicals. Any neural structure that emerged in response to a low steroid environment would be particularly susceptible to any major changes in steroid levels. If one half of the brain was more sensitive to steroids than the other, and then steroids were reintroduced, the small archaic differential between the hemispheres could over time be magnified greatly.

To illustrate this further, picture two sides of the archaic primate brain and let's say the left hemisphere had, for whatever reason, a 1% greater sensitivity to steroids than the right. This would have not been significant to the archaic primate in terms of structure or function. Maybe it was a minimally useful variation or maybe just the way it was. Then, some 70 million years ago, the steroid veil was slowly and progressively lifted as a result of a dietary switch to biochemically rich tropical fruit. The brains of primates slowly expanded, and, in the higher hominids, this expansion became exponential when the pineal loop took effect. Brain cells, which never would have emerged in a high steroid environment, now developed. Their structure and function depended on low steroid activity and any change in regime would of course change both structure and activity. With a change of diet, a degenerating pineal and the resulting imposition of a different steroid regime, the archaic difference in susceptibility to steroids led to one side of the brain being more adversely affected than the other.

Why did the side of the brain which was most affected by steroids become dominant? With the greater levels of steroids impinging differentially on the two sides of the brain, a split in function seems to have occurred. What emerged from this split were two different senses of self which could not coexist without confusion. For a time then, two selves may have existed side by side and this could have led to a host of strange effects, perhaps including an internal suspicion, the hearing of voices and much else reminiscent of the symptoms of schizophrenia today. The most dysfunctional side would be likely to experience the most distrust. Loss of function equates with a reducing ability to experience, to know and to understand, and with this loss comes a fear.

When we lose any perceptual sense, such as sight, the world becomes a more frightening place. For the left hemisphere's personality, losing some subtle senses engendered an underlying anxiety. Anxiety and fear are strong emotions that tend to mask all other mental functions. The more complex function of the right hemisphere became suppressed by a left hemisphere personality that kept, and still keeps, itself busy by generating a continuous internal dialogue stemming from anxiety. In this way it manages to shut out what it regards as the more baffling perceptions arising in the right hemisphere. Furthermore, it actively resists experiences, which it cannot fully understand or categorise, because these threaten its own sense of self.

Today we know that our human brains show a markedly lateralised function. There are small differences between the two sides of reptilian and other mammalian brains but the difference in humans is of a different order entirely. Such a difference in function between the left and right hemispheres is another strange and unique feature of the human brain that has never before been adequately explained.

Significant evidence is emerging to support this scenario. Professor Alan Snyder (Director, Centre for the Mind, Australia), Dr Darold Treffert (University of Wisconsin Medical School) and Professor Vilayanur Ramachandran (Director of the Centre for Brain and Cognition, California) have all found that the dominant side of the brain has a surprising degree of dysfunction (as we will see in the next chapter). We believe that more research will establish beyond doubt that the left hemisphere has lost the ability to run the body's biochemical functions at an optimum level, and that this loss of function, stemming from its greater susceptibility to steroid damage, has had a disproportionate affect on the whole body/mind system.

The rapid expansion of the human brain, the most delicate and complex mechanism that nature has ever constructed, ceased at some time around 200,000 years ago. Since then, what evidence there is suggests that our whole brain system has been shrinking. Leaving the forest and losing the complex cocktail of powerful steroid modifying chemicals that were permanently present during 70 million years of evolution has had a devastating affect on human neural structure and function. Without the optimum hormonal environment, our brains can no longer develop to their full potential and this appears to be particularly the case for our left hemispheres.

THE IMPLICATIONS OF AGRICULTURE ~ψ~

Some twelve thousand years ago there was another major change in the history of mankind which was to have profound implications. The establishment of agriculture shifted the basic diet from one of wild foods to one in which grain formed the staple. A further factor was the way the grain was processed before consumption. Roasting, baking and boiling heat-denatured the delicate structure of the living foods causing a decline in the quality of nutrition.

There is increasing speculation and some evidence that a catastrophic event fuelled these changes. Whole new patterns of existence emerged – there was a shift to certain cereal crops, which interestingly derive from mountain plants. (Could this, as myths from all corners of the world tell us, have been due to widespread lowland flooding?) Rice, wheat, barley and the edible tubers were originally high altitude crops and along with these new foods came the means to make them more edible. Cooking was already used as a tool to make, particularly meat, more palatable but at this time there was a much bigger move towards this way of food preparation.

Biochemically, the change to cooking is massive. It is a process that has never been a part of biological evolution. Heating food to high, burning levels is a trait that has emerged extremely recently in the three billion years of life on earth. All biology operates within a fairly narrow range of temperatures. It is chemically and enzymically orientated. The flow between species functions within these limits. To bring heat denaturing into the equation, particularly within a complex organism that had established a balance with its environment over aeons of time, was bound to have some repercussions.

An important study, completed over half a century ago, sheds light on the question of cooked versus raw food. In 1950, Dr Mananore Kuratsune, head of the Medical Department of the University of Kyushu, investigated the diet that was given to the prisoners Japan took during the last world war. This diet, consisting of around 800 calories per day per 70 kg of body weight, was well under a half of the daily minimum that is recommended to maintain health. The good doctor and his wife ate a raw version of this diet and both remained healthy but when they switched to eating the diet in cooked form all the symptoms of malnutrition that devastated the inmates of the Japanese camps rapidly showed themselves. These included oedema, vitamin deficiency and physical collapse. They were forced to abandon the experiment because they became so ill. They proved however that what was regarded as a grossly inadequate diet sustained them when eaten raw but did drastic damage when cooked.

Nutritional research has more recently given us endless examples of the detrimental effects of consuming heat-processed foods. Phosphatases (enzymes that break down phosphorus containing compounds) in milk are destroyed by pasteurisation rendering the calcium insoluble and making milk constipating. Heated unsaturated oils like safflower have been found to contain numerous poisonous compounds – some are powerful oxidisers and others are carcinogens. In May 2002, a worldwide alert was issued after scientists announced that much of the food we eat contains significant levels of acrylamide, a chemical known to cause cancer, affect fertility and damage the nervous system. Acrylamide occurs in fried, baked and processed foods ranging from biscuits, bread and crisps to chips and meat. It is formed in the cooking process and longer cooking means more acrylamide is formed. Amongst the products tested in the British study (crisps, crackers, processed breakfast cereals and chipped supermarket potatoes) some had levels of acrylamide 1,280 times higher than international safety limits. It is particularly nasty stuff; as a genotoxic carcinogen it has no safe dose. As 30 to 40 percent of cancers are caused by diet, acrylamide (and foods cooked in a way that encourages its formation) could possible emerge as one of the major causes of this devastating illness so prevalent in our world today.

Perhaps Paul Kouchakoff of the Institute of Clinical Chemistry at Lausanne provided the most striking evidence that cooked foods are unsuitable for our diet. He found that when we eat cooked foods white blood cells (leukocytes) rush to the blood vessels supplying the intestines to defend the body against the perceived threat of invasion. This effect was termed digestive leucocytosis and thought to be a normal reaction to the ingestion of all food but then it was found that raw food doesn't trigger this reaction.

At the same time, other research on healthy food has constantly reinforced the benefits of eating fruit and vegetables. Such foods are rich in biologically active phytochemicals that not only provide us with energy and raw materials for construction but, as we have already noted, have been found to protect us from cancer. They may do this in a number of ways. For instance, supforaphane, a phytochemical found in broccoli, can activate enzymes in our cells that removes carcinogens before they can cause any harm. Flavonoids, found in citrus fruits and berries, keep cancer causing hormones from latching on to cells in the first place. Genistein, found in soya beans, kills tumours by preventing the formation of capillaries needed to nourish them. Indoles, found in cabbages and Brussels sprouts, increase immune activity and make it easier for the body to excrete toxins, and saponins, found in beans and lentils, may prevent cancer cells from multiplying. We are only just beginning to discover the benefits of these substances and as they are extremely numerous (it has been estimated that there are 10,000 different phytochemicals in tomatoes alone) there is much more to find out.

Generally we do know however that diets high in fruit, vegetables, grains and legumes appear to reduce the risk of a number of diseases, including cancer, heart disease, diabetes and high blood pressure. Some of these phytochemicals may survive the cooking process (Genistein for example is present in such processed products as tofu and miso soup) but the antioxidants that are also present in a fruit and vegetable-based diet and which also protect cells from cancerous growth are more vulnerable to heating. It seems therefore that to extract the most health benefit from our food we should not only eat a diet of predominantly fruit and vegetables but that we should eat it with little or ideally no cooking.

There is a large body of scientific and anecdotal evidence that extols the virtues of raw foods. For example, they raise micro-electric potentials throughout the body, boost metabolic functions, increase resistance to illness and the speed of healing, and enhance the transportation of nutrients through the capillaries which, in itself, aids the removal of toxins from the whole body system. Leslie and Susannah Kenton in 'Raw Energy' summarise the benefits of raw food concisely.

'Over quite a short time an all raw, or nearly all raw diet, does several things. It eliminates accumulated wastes and toxins. It restores optimal sodium/potassium and

acid/alkali balance. It supplies and/or restores the level of nutrients essential for optimum cell function. It increases the efficiency with which cells take up oxygen, necessary for the release of energy with which to carry out their multifarious activities.'

They also highlight, with many examples, how raw diets have helped to heal numerous diseases including arthritis, diabetes and cancer, and how they help to retard ageing too. This all adds weight to the assertion that building and running our systems on a less than ideal nutrition has had far reaching consequences.

Cooking therefore represents one further step away from the ancestral diet that nurtured our brain growth. This new change accelerated the degeneration that had been slowly taking place over the preceding 200,000 years. A system, which was already on a downward slope, now constructed itself out of materials ever lacking in the biochemistry needed for healthy functioning.

Evidence shows that whenever a culture crosses over from a wild food diet to an agrarian one there is a decline in health – skeletons are smaller and show more disease. If such a change in diet can so affect the skeleton, what are the implications for the construction of the much more delicate and sensitive brain?

A psychological feedback loop may have made the situation even worse. One effect of left hemisphere degeneration is the need for our fear-based dominant side to cling to what becomes familiar. This can include detrimental cultural approaches. When 'advances' like eating foods in a certain way becomes part of a culture it is difficult to change again. What we are familiar with is what we do. When a tribe establishes a diet of cooked tubers and hunts for specific animals this becomes part of their identity. This psychological component of a culture is wholly different from, for example, an antelope eating grass every day (the food it was evolved to eat). When there are no tubers or animals left to hunt, tribes may end up half-starving rather than eating something that is culturally unfamiliar. Some tribes, despite living in warm environments that could support many different rich food sources, cannot comprehend changing their way of life or giving up their herds.

And could we in the 'west' give up our dependency on what we regard as a standard diet and live on something which maybe more nutritious? We know how difficult it is to get our children to eat fruit and vegetables after they have become fixated on fatty and sweet processed foods. It would be hard for our farmers to give up their grain and livestock industry to grow food which would form the basis of a fruit/vegetable diet, and difficult for consumers to make wholesale changes too. Our familiar biochemical intake is known to be detrimental to our health but would it be possible for us to make a radical change even if it was proved beyond doubt that our diet is as bad as we have suggested?

JUNK FOOD AND DEGENERATION ~ψ~

Today our diet comprises not only cooked food but also cooked, processed and refined foods, which are far from natural. This takes the situation an absurd step further. The problem worsened with the onset of the agrarian revolution but the junk food dimension is just the final insult that is leading to ever increasing degeneration.

Weston Price, a dentist and nutritional pioneer, in a huge study made in the 1930s, documented case after case of communities which suffered from degenerative disease after a change was made from a traditional 'native' diet to one based on white flour, canned foods and biscuits. From isolated communities in the Hebrides to the Maoris of New Zealand, he found a massive increase in tooth decay, dental arch malformation and a decrease in immunity to disease when the change was made to pre-processed foods. He also implied that the consciousness of the people changes with diet too. He reported on the happiness of the Ugandan people prior to modernising elements of 'civilisation' – a report which has acquired all the more poignancy in the light of the recent history of that troubled country. Of the Torres Islanders he said 'it would be difficult to find a more happy and contented people. Their home life reaches a very high ideal and among them there is practically no crime.' After western food was imported into these islands, the children born to mothers eating these new foods began to show gross deformities of their dental arches, a narrowing of their faces, pinched nostrils and a crowding of their teeth. These 'deformities' are so common in the west that they are regarded as natural and normal. No-one questions why so many of us have crowded and overlapping teeth and why it is common to find nostrils so narrow that one is forced to breath through the mouth. In their native state the Torres Islanders had exceedingly little disease and within a thirteen year study period not a single case of malignancy. However when they changed from their traditional diet of bananas, papaya, taro, plums and seafood to white man's food they began to suffer a loss of immunity to dental caries and in succeeding generations showed a marked change in facial structure and a lowering of resistance to disease.

Price recorded this pattern of degeneration across the globe in traditional communities newly infected by a modern western diet. He investigated communities in the high mountain in Switzerland, Gaelic folk on the Isle of Harris, Indians in Alaska and throughout America, Pygmies in the Congo, Aborigines in Australia, Maoris in New Zealand and many Polynesian Islanders. His evidence is overwhelming and damming but, unfortunately, comprehensively ignored.

Abram Hoffer in his introduction to the fifth edition of Price's 'Nutrition and Degeneration' notes that:

'Recent intergenerational research in animals and people has shown that on a uniformly poor diet, the offspring of each generation deteriorates more and more'. He suspects that *'many of the people with psychiatric disorders today, the addicts, the high degree of violence, the tremendous amount of depressions and tension states, and the great number of physical degenerations such as diabetes, arthritis etc. are the modern manifestations of this continuing degeneration'.*

Even our ability to hear properly is affected by diet. Finnish researchers have found that people on a low fat, low cholesterol diet had better circulation to the ears, and consequently better hearing. Another large study conducted at the West Virginia University School of medicine concluded that hearing improved with a diet rich in fruit, vegetables and whole grain cereals. With many patients, dizziness cleared up promptly and the sensation of pressure in the ears and head was quickly relieved.

If our thesis were correct we would expect to see increasing evidence of degeneration all around us in response to our increasingly refined, processed and sweet foods. And indeed today we are seeing widespread problems. Our systems, despite being powerful and flexible, are beginning to crack under the strain. 'Failure to thrive' syndrome is becoming more common in the western world plus chronic fatigue, MS, cancers, immune problems, deprivation dwarfism and a re-emergence of tuberculosis. Some infants even have to be force fed just to get them to live. This phenomenon has only emerged in the last 20 years. Could it be due to second and third generations raised on junk food? By medical intervention such cases can be successfully treated but what does all this say about the wider picture of our nutrition and disease? Kids eating junk food are unknowingly harming themselves, as are their parents, eating highly processed convenience meals. And mothers eating a junk food diet could inadvertently be harming their yet unborn children.

Autism levels are reported to have increased tenfold in the last decade (BBC News Online: Health (27.02.2001). Links have been tentatively made to the MMR vaccination but if, as we have pointed out earlier, autism is in some way caused by steroid levels the figures are not surprising. Human beings today are exposed to thousands of chemicals in the air, water and food. We are inadvertently being dosed with a multitude of drugs from food additives, pesticides, fertilisers, industrial effluents, paints and plastics. Many of these products are completely foreign to man's biochemical system but some are very like our steroids. What effect are they having? Can the decrease in the age of puberty be linked to environmental steroids?

In our 'western culture' and particularly in populations where there has been a recent change from traditional diets to a high fat, high sugar 'modern diet', the age of puberty has been reduced by some 20% in three generations. This is a massive change. It demonstrates the plasticity of the human system and how it can radically change in response to changes in steroid levels. In girls particularly, the change can also be linked to an increasing fat intake. Oestrogen (a steroid hormone which initiates female puberty) is produced in layers of fat as well as in the ovaries. Increasing fat means more oestrogen is produced thus initiating puberty at an earlier age. Increased levels of this hormone will also affect the hormonal environment in the uterus when they are pregnant and this, as we have seen already, will affect their offspring. The generational cycle continues forever onwards and, if it is fuelled by inadequate and damaging nutrition, the cycle will spiral downwards.

The age of puberty may also be affected by our artificially increased day length. Bright electric lights and television suppress melatonin production. Roberto Salti, in a study conducted at the University of Florence, found that children who abstained from watching TV for only a week significantly increased their melatonin levels. It actually increased by about 30%. If future studies show that this suppression of melatonin is linked with earlier puberty it will almost prove our hypothesis in reverse. Less melatonin leading to a shorter juvenile period is exactly the opposite of what we are proposing happened during human development in the rainforest. If more melatonin led to a longer juvenile period and bigger brains, what will be the effect of this modern reversal?

The medical and scientific communities largely assume that increasingly early puberty has no consequences but is this laissez-faire attitude justified? A shorter juvenile period means a shorter window for neural development, and this reduction may even produce a mild retardation. Eunuchs apparently have a greater longevity than testosterone-pumping males, so it seems that lower steroid levels do have some positive effects. Perhaps a late puberty would allow the neural and neuroendocrine systems to develop a little more which would increase both the life span and the health of the individual.

Conversely, a retarded neural-endocrine system, due to the loss of perhaps two years of the juvenile period, could lead to all sorts of problems, including a less efficient immune system. And in females it would also lead to a different hormonal regime in their uterus with knock on effects in their children. It may be significant too that melatonin is present in breast milk but not the artificial replacements. Do children fed on breast milk reach puberty later? We haven't seen any data on this but we suspect it to be so. Research indicates that babies do not make much of their own melatonin for the first three months of their lives so the melatonin in breast milk may be particularly important at this crucial time of neural development. There is evidence that breast-feeding for longer improves intelligence and immune function.

Alzheimer's disease is another degenerative disease that is becoming uncomfortably common. Work at Washington University School of Medicine has recently identified a precursor of the disease they term 'mild cognitive impairment' characterised by repeated lapses in short-term memory. It appears then that Alzheimer's may be beginning much earlier in a person's life than previously thought. As usual when such diseases are identified, the medical/pharmaceutical establishment looks for a drug 'cure' and not at the wider picture of nutrition. But in a parallel area of research, carried out at the University of Alabama, it has been found that soy protein may reduce the risk of not only Alzheimer's but also heart disease and cancer. The natural chemicals within Soya beans mimic the structure of human oestrogen but actually suppress the steroid action by perhaps out competing and blocking its activity. This appears to be yet another instance of a chemical within a food plant preventing disease by suppression of our (unnaturally) high levels of steroid hormones.

It could be argued that making human populations from such terrible construction materials cannot but lead to degenerative problems. If someone set out to devise the worst possible diet with which to build a huge-brained tropical primate, a junk food diet of white bread, crisps, pies, chips, jam, toffee bars, biscuits, sweet fizzy drinks and burgers would be it. The cruel twist is that we seem addicted to these foods. In fact recent work suggests that eating fat rich foods stimulate opioids in the brain. It has even been humorously stated that hamburgers are as addictive as heroin. We seem to have lost the sensitivity to know what is best for us.

Wild animals do not appear to be subjected to the degenerative processes affecting modern people. Price ascribed this to an animal instinct in the matter of food selection. There are many instances in which animals deliberately seek out and consume mineral rich deposits necessary for the maintenance of their health. It is possible that we humans have lost, through disuse, our instinct for consciously recognising what our bodies really need. The only hunger of which we are now aware is the hunger for energy to keep us warm and active. In general we stop eating when an adequate amount of energy has been provided (or when we are just too full to eat any more) whether or not the bodybuilding and repairing materials have been included in the food. And of course there are also cravings for foods which seem to provide a 'comfort factor' – usually sweet things which give a quick hit of energy. (It is possible however that some addictive foods, such as chocolate, may be providing chemicals that we instinctively know our brains need. Perhaps they act as substitutes for similar chemicals, which would have been provided by a diet rich in tropical fruit chemicals.)

Much of the evidence we have presented in this section shows that degeneration and disease is linked to increasingly poor nutrition. The pain and heartache of our nearest and dearest dying from what seems to be an epidemic of cancers may be just one of the sharpest symptoms of the loss of an archaic diet rich in health promoting biochemistry.

SUMMARY ~ψ~

We have come a long way in this chapter and have been very ambitious. We have attempted to explain just how human beings attained such large brains, when and where it occurred and what has happened since that time of unprecedented neural expansion. In doing so we have revised the biological foundation of our species.

We propose that a specialist fruit diet flooded our systems with chemicals that boosted neurotransmitter activity and reduced the activity of our bodies' steroids. This combination would have had some slight effect on the developing child in the uterus because a change in steroid activity changes how DNA is read and thus what is built. This change may have affected, amongst other things, the functioning of the pineal. We suspect that, at a critical juncture in our history, the pineal became more active, increasingly enhancing the production of melatonin and monoamine oxidase inhibitors. These further boost neurotransmitter activity and reduce steroid activity. Thus a unique and increasingly active loop was established that lead to a rapid expansion of our brains.

With the loss of our original forest habitat, a change of diet was imposed upon us. We survived because we were the cleverest hominid around, but the loss of the forest fruit biochemistry had an increasingly negative affect, as our increasingly damaged left hemispheres became dominant. During the last 200,000 years or so this dominant hemisphere may have undergone some change: the language centre may have developed to some degree but overall the change has been deleterious. In particular, our sense of self has become distorted. We have become fearful, violent and disconnected from the environment that supports us. The quality of our consciousness has changed.

CHAPTER FOUR

~ψ~

Consciousness

In this chapter we shed more light on the two sides of the self. A hidden side that possesses enhanced function is revealed in certain cases of hemispherectomy, schizophrenia, multiple personality disorder, through the use of hypnosis and in experiments in which one side of the brain has been anaesthetised. We propose that the right hemisphere has abilities that can only emerge when left hemisphere dominance is lifted. The sense of self, which is liberated during the fleeting moments when left hemisphere activity is reduced or by-passed, relates to what is often interpreted as religious experience. Much evidence is presented that indicates that the left hemisphere in humans is less functional than the right. Over the course of human history, this has had enormous and tragic consequences.

It is generally assumed that the brain has something to do with consciousness. Brain damage and drug use can radically change our sense of self whereas operations like heart transplants do not tend to. The general tenant of the scientific model proposes that consciousness is a result of complex biochemical interactions that we do not fully understand. This may be the case, though it is more probable that the picture contains levels of far greater subtlety. However, for the time being, let us assume that cellular and sub-cellular structure within the brain facilitate the sense of self and, building on this, it follows that any change in structure is liable to change this sense of self.

A musical analogy may be helpful here: imagine that a violin represents the structure of the brain and the music the sense of self. If the tension in the strings is altered a different sound will be produced. And if the wood from which the violin is constructed is changed, a different quality of sound will be generated. There is a tangible connection between the violin and the sound it produces; yet one is not the other. The structure of the brain is not the sense of self but there is a direct correlation. The more complex the instrument the more subtle the sound will be; and small changes to the structure will have a correspondingly greater affect.

We have already noted that our whole 'brain/sense of self system' was once built from the best quality materials. The tropical fruit diet represented the equivalent of the very best violin varnish and the finest of woods but with a change of diet the quality of construction materials declined. This would have affected both the structure of the brain and the function of the 'brain/sense of self system'. The sense of 'I' would have been changed.

According to our hypothesis, before 'the fall', our two brain hemispheres worked in tandem. They produced a single beautiful sound – a complete and connected sense of 'I'. But, the increasing activity of steroids in our system changed the structures that produced this unified sense. The structures and interfaces of the left hemisphere were altered to a *greater degree* than the right, and as a result it became comparatively less functional. Evidence of this has already been given – for example: LSD only elicits the typical perceptual response in the right hemisphere. Yet the two halves of the brain still share the same cranium. Both make a 'sound' that we usually perceive as a hybrid 'note'. We rarely experience two distinct senses of selves.

Most of us have however experienced, usually fleetingly, a greater sense of something. This may occur spontaneously, perhaps in the presence of great natural beauty, but more often as a response to mystical or religious practices. While such shifts are usually positive, they can sometimes undermine our 'normal' sense of self. If the perceptual change during these experiences is a result of a shift in balance from left to right, perhaps these uncomfortable feelings stem from some internal conflict between the two sides.

While the ancestral biochemical and structural changes to our neural system were slowly but progressively taking place, elements of confusion would have disrupted our original self of self. Before the 'fall', we would have felt more 'connected'. Greater perceptual function would have facilitated a different sense of self that would have linked us to our family group and our environment in a more profound way. The feelings of separateness and disconnection, and the ensuing symptoms of fear and anxiety which are endemic today, are fundamentally a reflection of the disconnection from not only our outer environs but also our whole, once integrated self.

If our two hemispheres were balanced and declined together at the same rate, we may not have noticed that anything was wrong – we wouldn't have this feeling of disconnection from something. But, because there is this disparity, we have an inbuilt, underlying knowing that there *is* 'something else'. It is like having a lost twin that we never consciously know about yet feel, at some deep level, connected to. The fact that our twin cerebral system is dominated by the least functional side has set up a bizarre problem: the self that is damaged dominates us and, because it is damaged, we don't realise the extent of the problem. It is stuck in its own distorted and limited version of reality and, as we will see later, does all it can to maintain this illusion.

THE BICAMERAL MIND ~ψ~

Julian Jaynes identified, from a detailed study of early writings, cultural archaeology and medical behaviourism, a state he termed the bicameral mind. He proposes that at one time human nature was split in two with an executive part called a god, and a follower part called a man. He argues that neither part was conscious in the way that we are today, and that consciousness arose when this bicamerality broke down under the pressures of social conflict and cultural change. His argument is fascinating and, whilst we have reached a different conclusion, his journey of discovery reveals much that is of interest to us here. He hypothesised that there was a critical period in our history during which there was greater access to right-sided function. The 'god side of self' is a manifestation of the right-sided self. The internal voice that comes from this side was equated with god because it has a directness and authority. It often tells the individual what to do. Thus Agamemnon 'had to obey' the 'cold command' of Zeus and Paul, the command of Jesus on the road to Damascus. Jaynes assumes that the voices heard during this historic transition were similar to the voices heard by not only present day schizophrenics but also by many 'normal' people when they are subjected to unusual or stressful events. Hearing these voices is taken by some to be a gift, perhaps even a divine gift. And they have an elemental power that seems hard to resist, even in extreme cases when for instance they command one to take one's own life. They take precedence over all logic and reason.

What we are seeing here, we believe, are the latter stages of the dual-self breakdown. The two sides, that once worked in harmony (or as one), diverged due to increasing left hemisphere damage. Before the left side established an overall functional dominance, there must have been a period during which right side function continued to work in parallel with the left. This period of the 'bicameral mind' is consistent with the confusion caused by two divergent selves working separately within a single cranium.

As the left hemisphere attained dominance, the final stages of right hemisphere conscious involvement was manifested, in part, by these divine voices. Our original, connectedness sense of self was reduced to the confusion of these internal voices and then on to the vague feelings of unsatisfactoriness that we feel today. We can see a response to this continuing loss of right-sided function – something Jaynes calls 'the slow withdrawing tide of divine voices' – in our attempts to make connection via religious rites and rituals. 'Prophets, poets, oracles, diviners, statue cults, mediums, astrologers, inspired saints, demon possession, tarot cards, ouija boards, popes and peyote are all residues of bicamerality'; and are indicative of a deep yearning for a reconnection to a lost state of being.

The decline of the right hemisphere control is illustrated in historical Greek culture by the declining role of their oracles. Accessing information from the Gods began simply at specific, maybe awesome, locations where any supplicants could still 'hear' bicameral voices in the sound of waves, water or wind. (Even today, we can feel something of this effect when 'worshipping' at a waterfall or in a cathedral.) The loss of this easy access resulted in special people being promoted to the position of 'oracle'. These mystical characters were consulted on all great matters of state, but over the thousand years of this tradition even their connection to the Gods tailed off. The voices became fitful, the possessed prophets erratic, interpretations impossible, and finally the days of the oracles came to an end.

An example of the lengths that people went to in order to 'hear' a divine message is described in detail, in A.D. 150, by Pausanius, a Roman traveller to Greece:

'After days of waiting and purification and omens and expectancy, he tells us how he was abruptly taken one night and bathed and anointed by two holy boys, then drank from Lethe's spring to forget who he was, then made to sip at the spring of Mnemosyne so as to remember later what was to be revealed. Then he was made to worship a secret image, then dressed in holy linen, girded with sacred ribbons and shod with special boots, and then only after more omens, if favourable, was finally inserted down an impassive ladder into the devout pit with its dark torrent where the divine message grew swiftly articulate.'

A very long process designed to loosen the stranglehold of left hemisphere dominance and allow the right to come out and play.

Parallel techniques are used in many rites, rituals and ceremonies to induce a sense of 'the other'. These often involve deliberate elements of stress such as dancing for days and nights with no sleep (the sun and eagle dances of North American Indian tradition) which pushes the human system beyond its normal limits. This too reduces the stranglehold of left hemisphere dominance and allows space for the sense of deeper connection to manifest.

LET'S SING ABOUT IT ~ψ~

The cultural decline of Greek oracles mirrored a similar change in the fortunes of the Muses. Jaynes believes that poetry is founded in the rhythmic speech of the bicameral mind. As the bicameral mind broke down, those who retained 'the gift' either became prophets or poets (or both) relating the words of the Gods to their community. As left hemisphere dominance increased, the poets, in order to access their muse, had to *learn* to do it and, as this became more difficult, they resorted to conjuring up states of ecstatic possession. By the end of the first millennium BC access had been mostly lost, and poetry became a *creative art*. Poets sculpted their own words in laborious imitation of previous divine utterances.

Unlike the oracles, poets did not disappear; their craft changed from one in which messages were received and directly passed on, to something that is in part a nostalgic search for the absolute. Some poets have however, even in these latter days, accessed states recalling those of over two thousand years earlier. The most celebrated of these was William Blake who was visited by extraordinary visions and auditory hallucinations that could persist for days at a time.

The phenomenon of hearing the voice of god as rhythmic poetry was once widespread. The epics of Greek culture were heard and spoken in verse form, as were the Vedas – the oldest Indian writings. Early Arabic people called their poets *sha'ir* – a word which meant 'endowed with knowledge by the spirits', and it was the metered form that authenticated the divine origins of their recitations. Hebrew prophets too relayed their messages from God in verse. Similar instances still occur; speaking in tongues tends to be rhythmic, and, in some cases of spontaneous possession, utterances are delivered in meter too. As noted in Chapter One, A.W. experienced of a day in which all his thoughts were in verse.

Early poetry was very close to song. In ancient Greece, words and phrases were not stressed by vocal intensity but by changing pitch – auditory ornamentation that gave variety and beauty to recitations. It is possible that this style of delivery tapped directly into the right hemisphere of the listener.

Singing is primarily a function of the right hemisphere. It is common medical knowledge that patients who have lost the power of speech, via cerebral haemorrhage in the left hemisphere, can still sing. Even patients who have had their entire left hemisphere removed, and can hardly speak, can not only sing but also clearly enunciate the words of the song.

There is a procedure, called the Wada Test, that is performed in hospitals to find out a person's cerebral dominance. Sodium amytal is injected into the carotid artery on one side, putting the corresponding hemisphere under heavy sedation, and leaving the other awake and alert. When the left hemisphere is knocked out, speech is also, but not the ability to sing. Surgery has shown that the specific area of the right hemisphere, which is engaged in musical function, is the right anterior temporal lobe. Patients who have had this area removed find it very difficult to distinguish one melody from another. They cannot recognise familiar tunes and, when asked to hum along with a melody, generally sing the wrong notes – they may end up just tapping out the rhythm. Patients that have had the anterior temporal lobe removed from the left hemisphere have no such inability.

The function of singing incorporates speech and vocabulary plus the added musical elements too. It appears then that the left hemisphere is not able to deal with this level of complexity. This could suggest we all have a latent musicality that is, to a greater or lesser extent, blocked by left hemisphere dominance. It is interesting to note too that 'singing words' is a recognised therapy for stammering. Singing provides a way of engaging the right hemisphere's speech function when the left has a problem with its so-called specialist adaptation.

It is significant that music, particularly when it is combined with rhythmical dance, can induce altered states of consciousness. Dancing to music with a strong beat has been used in all cultures, and often in the context of spiritual ritual. From tribal dances and whirling dervishes to our present day trance dance raves, music combined with body movement has precipitated emotional release and a return to wholeness. The therapeutic value of music lies in its ability to conjure up emotions from ecstasy to melancholy – and this is right hemisphere territory.

It is usually accepted that speech is the specialist function, par excellence, of the left hemisphere but this is now open to question. Parts of the left hemisphere however are more developed than the corresponding sections in the right. Wernicke's area in particular is thick with large, widely spaced cells that are indicative of considerable neurological connections. This part, together with Broca's and the supplementary motor area, is associated with speech. Any large destruction of Wernicke's area will produce a loss of meaningful speech, but the permanency of this loss depends on the age of the individual suffering the damage. In a child, a major lesion on Wernicke's area in the left hemisphere, or on the underlying thalamus that connects it to the brainstem, will result in transfer of the whole speech area to the right hemisphere. Thus the 'so called' speechless right hemisphere can under certain conditions become a language hemisphere just like the left. In fact it always was a language hemisphere, for we can see that stroke patients who have suffered haemorrhages within the left side of their cortex, despite losing their speech faculty, can still understand verbal dialogue.

Language and speech is taken by many to be the crowning glory of human achievement, and it was always thought to be the provenance of the clever, dominant left hemisphere. We can now see that the right hemisphere possesses these abilities too. While Wernicke's area in the left hemisphere does look well developed, this may be nothing more than a response to continued use. Even a muscle will become larger if it is used more, and speech is one activity we seem to constantly engage in. The left may be more damaged than the right but the fact that it has assumed dominance means that it does what it can do for much of the time.

Speech is one of the main functions that the left hemisphere appears to excel in but, without the support of the right hemisphere, it even has problems with this. Has the left hemisphere then any abilities that are not possessed by the right? From the evidence given above, it appears that the right can assume all the functions of the left if the change to right function is made early enough in the individual's life. But the left does not seem to be able to manage so well without the right hemisphere.

The Written Word and Cerebral Dominance

Although Jaynes didn't make the connection in his work on bicamerality, the need for reading and writing can also be interpreted as a response to a mental failing that included declining memory and reduced access to direct intuitive knowledge. The movement towards writing started with simple representations of an object (pictographs) – the earliest we know about date from around 40,000 years ago. By 15,000 years ago people were marking sticks and bones with information on the sun's cycle, crops and the movement of animals. Ideograms, symbols that represent an idea or word, developed some 10,000 years later and are still used by the Chinese and Japanese today. These, together with the sophisticated Semitic scripts, require more from the reader than the simplified Latin scripts of the Greeks and Romans. The Semitic scripts (Hebrew and Aramaic) are somewhat similar to shorthand. Only consonants are included so that the reader has to deduce the vowels from the context. This gives these languages an intricacy that allows the conveyance of greater layers of meaning and subtlety of nuance. A picture is indeed worth a thousand words. As Robert Ornstein, author of 'The Right Mind', remarks, 'this makes Hebrew and Aramaic redolent with resonance, almost poetry in every phrase, as the root meanings resound to the listener'.

It is no accident that scripts that include vowels are almost all written towards the right whilst the languages without vowels are written towards the left. When reading towards the right, as you are now, the images enter the right visual field (of each eye) first and are thus initially processed by the left hemisphere. The reverse is the case for scripts (and hieroglyphs) read towards the left. We know that the left hemisphere cannot handle context. Without a right hemisphere sophisticated jokes, for instance, are impossible to understand, as they require a multi-layered level of comprehension. The shift then from complex writing scripts to simplified ones, goes hand in hand with the historic period of increasing left hemisphere dominance, as identified by Julian Jaynes. It is likely that this increasing shift made it progressively harder for readers to use scripts like Hebrew and Aramaic.

Recently it has even been proposed that the invention of writing, particularly alphabetic writing, actually rewired the brains of those who used this culture-changing tool. Leonard Schlain has argued that literacy reinforced the brain's linear, abstract left hemisphere at the expense of the holistic, iconic right one, and this may have initiated cultural changes such as the decline in the importance of the 'goddess' and the rise of patriarchal religions. Rather than creating the problem or rewiring the brain, we believe that literacy, based on linear texts, exacerbated an already dysfunctional situation. This helped to reinforce the left's dominant reality of word-based beliefs and concepts at the expense of the right's non-verbal experiential reality.

HEMISPHERECTOMY ~ψ~

Sometimes, as in cases of extreme epilepsy, it becomes medically expedient to surgically remove one side of the brain. Although these patients remain to some degree brain impaired, from them we can glean useful insights into the respective function of each separate hemisphere. Having only one hemisphere takes away the ambiguity of any hybrid sense of self.

We can also see from cases of stroke how one side functions with reduced interference from the other. Stroke damage, for some reason, is more common in the left hemisphere, but to begin with lets look at what emerges when the right is damaged or has been removed entirely.

Right hemisphere removal

Cases of patients who have had their right hemisphere removed are particularly interesting for they give us the best indication so far available as to what the left brain can and cannot do. And it appears that the left has more dysfunction than expected. The left after all is the dominant hemisphere and has been regarded by many as the more developed half.

Every case is different of course, and every individual will have had different levels of illness and disabilities before their operations, but nevertheless clear patterns can be seen.

Patients with no right hemisphere become very much speech-based but do not have much recall. Memory becomes very poor. The remaining brain has the ability to work things out but in a very simplistic, linear way, processing one bit of information at a time. It still retains the ability to label and categorise, but an isolated left self appears to be very mechanistic, lacking emotion, and even recognition is much reduced. There is an inability to distinguish male and female voices, and the voice itself tends to be delivered in a monotone.

Emotionally, 'left hemisphere people' may have indifferent or even positive outlooks on life, yet these states often seem unrealistic and disconnected. There can be a fragile level of contentedness, and a desperate need for familiarity and routine, which if challenged can rapidly lead to high levels of anxiety. One 27 year-old male was reported as emotionally unstable, inconsistent and lacking in perseverance, but was not greatly disturbed by his physical handicap. In another case, a 45 year-old man displayed a narrow range of associations, little originality or imagination, and no ability for deep thinking or for adequate self- reflection.

Commonly 'left hemisphere people' have great difficulty responding to emotion in appropriate ways. This inability is connected to their lack of comprehension of any verbal messages beyond the literal. They do not understand any implied undercurrent in dialogue or stories, and can never understand jokes. They miss almost entirely the context in which their own words or the words of others are framed and so can never appreciate any subtlety in communication. In an experiment designed to test the understanding of the metaphor contained in the phrase 'a heavy heart can make a difference', it was found that patients with an unsupported left hemisphere most often chose a picture (a person carrying a large red heart) that represented a literal interpretation. In contrast, those patients with intact right hemispheres but damaged left ones usually chose the metaphorical picture (a person crying) and rejected the literal ones as funny and absurd.

Damage to the right hemisphere interferes with spatial orientation. Patients without a right hemisphere can lose the ability to know where they are entirely. In one case an individual couldn't find his way out of his own house without counting the doors. Another was able to name their local hospital and the appropriate ward number but when they actually arrived at a consulting room they had been to many times before, there was bewilderment. They had no recollection of the room.

In tests using mazes, patients with a damaged or removed right hemisphere find learning pathways almost impossible. In contrast, patients with equivalent left hemisphere limitation appear to have no such problem. Tests matching geometric shapes are also extremely difficult for subjects without functional right hemispheres but damage to the left makes no difference to this ability either.

Failure to recognise individual faces and to distinguish differences between facial expressions is yet another category of ability that is lost when the right hemisphere is damaged or removed. Patients may not even recognise their own face or the faces of their close family until they are specifically named. And they have great difficulty distinguishing any emotion too, be it anger or joy, in both faces and vocal intonation. Tests using tape-recorded sounds of such things as coughing, snoring and waves breaking on the shore are also hard for the left hemisphere person to identify.

There appears to be some lack of connection between vision and interpretation. A left hemisphere person will often fail to notice missing elements on simple pictures – animals without tails, figures without hands and feet etc. And visual recognition of the season can be problematic too. If presented with the sights of snowdrifts and leafless trees, a person lacking a right hemisphere will not be able to make the connection to winter. Yet given the information that it is January, the deduction can be made with ease.

One of the strangest effects of right hemisphere damage is left side 'neglect'. In this situation, patients ignore the entire left side of space. They may shave only the right side of their face, eat food from the right side of their plate and when dressing may forget to put their left leg in their trousers. In extreme cases they actually deny that the left side of their body exists.

To summarise, we can say that the left hemisphere can classify information but, without the underlying support of the right, it displays a great inability to perform many basic functions. These include memory, interpretation of emotion, spatial awareness and interpretation of elements within the visual field. All these are complex, multi-leveled abilities.

Left hemisphere removal

There are fewer documented cases of left hemisphere removal than right and hence correspondingly less information on 'naked' right hemisphere function. In the past there was understandable resistance to this procedure as the left hemisphere was regarded as the dominant and most important part of the brain. It was thought to remove the left would lead to catastrophic results. Yet operations have been successful, and from them we can gain some insight into right hemisphere function.

The most obvious negative response is a reduction of vocabulary. Speech is constructed of short phrases or isolated words, and gesture and mime are often used in preference. However intonation and ability to recognise intonation is better than in the normal 'dual-hemisphere' state. Hearing non-verbal sounds, such as crashing surf, is better too, as is musical recognition and response. It seems that the lack of competition or suppression from the left brain improves the performance of these tasks. Interestingly too, when the left hemisphere is damaged or removed, there is no equivalent neglect of the right side of the body.

Patients retain their grasp on spatial relationships, can recognise faces (though names may be lost) and can recognise and respond appropriately in emotional situations. To summarise – they have a correct perception of reality but have real difficulty communicating with words and language. Such a complex of conditions is displayed in the following representative case.

Peggy Gott, a colleague of Roger Sperry's At Cal Tech, examined various hemispherectomy patients and commented in detail on a girl who, at ten years of age, had the left half of her brain removed because of a malignant tumour. She retained some ability to create expressive speech, however she was limited to single words or short phrases. She could sing and was good at it but had some difficulty with numbers – although she could count up to 30 and write numbers to 10, she had a problem naming them. She also lost the ability to assign a verbal name to a visual symbol. However, personality characteristics such as humour, boredom, love and frustration appeared to be entirely unaffected by the surgery – her parents noted little difference in this area at all.

In contrast to the left hemisphere then, the right emerges as possibly more functional than expected. Certainly removing the left does not have the severely negative consequences that were predicted by those that regarded the left as the major hemisphere.

Right hemisphere insight

According to our hypothesis, the right hemisphere function that we do see in such cases does not represent its full ability, for the right brain has still been subject to a lifetime of inadequate biochemistry and left hemisphere suppression. If these factors could be somehow corrected what results would there be?

Moments of insight, direct knowing, and peak experiences of joy suggest that we all possess a more amazing body/mind system than we generally realise. In our more usual state of consciousness, our right brain processes are interpreted through our left brain filter, which only seems able to deal with codified and simpler linear systems. This tends to dampen our everyday experience but, just sometimes, we get glimpses of astounding non-linear function; something more powerful takes over. Could this be unhindered right brain function?

The greatest scientific insights have not come in the laboratory or from the study but from what has be termed the 'bed, bath or bus'. Whilst the Russian chemist Dmitri

Mendeleyev was a Professor at Saint Petersburg in the 1860s, he engaged in a struggle, without much initial success, to find some order in the chemical elements according to their atomic weights. His intense waking efforts were stuck when he saw in a dream a table in which all the elements fell into place. From the notes he made upon awaking, and with only one minor change, Mendeleyev established the Periodic Table of Elements which graces the wall of so many science classrooms today. But the story did not end there. Based on his dream table, he predicted the existence of three 'non-existent' elements, all of which were discovered within the next fifteen years.

There are many other famous instances in which states of dreaming or revelry out perform normal waking abilities. Gauss, referring to an arithmetical theorem which had been troubling him for years, wrote how 'like a sudden flash of lightening, the riddle happened to be solved'. Helmholtz's insights 'crept quietly into his thinking' often while walking over wooded hills in sunny weather, and Einstein's greatest ideas came suddenly while shaving or during dreams. Such occurrences are common. Most of us have had similar experiences. It is only when we stop thinking where, for instance, we have left our car keys that their whereabouts comes to us. The release of the controlling aspect of linear function that occurs during states of revelry and dreams allows the right brain to function more freely – in this state creative ideas seem to appear from nowhere and we can access a much more detailed part of our memory too.

Though the idea of right hemisphere activity and creativity has been around for a long time, this has only recently been experimentally established. Within the scientific community popularist ideas of a left/right split for linear rationality and intuitive creativity was for a long time dismissed as overly simplistic but there is now positive evidence that increased activation of the right hemisphere is associated with enhanced creativity. Psychologists Brad Folley and Sohee Park from the Vanderbilt Kennedy Center for Research on Human Development have found that individuals with schizotypal personalities – people characterized by odd behaviour and language but who are not psychotic or schizophrenic – are more creative than either normal or fully schizophrenic individuals. These types rely more heavily on the right sides of their brains than the general population to access their creativity.

Folley and Park conducted two experiments to compare the creative thinking processes of schizotypes, schizophrenics and normal control subjects. In the first, research subjects were shown a variety of household objects and asked them to make up new functions for them. The results showed that the schizotypes were better able to creatively suggest new uses for the objects; the schizophrenics and average subjects performed similarly to one another. In the second experiment, the three groups again were asked to identify new uses for everyday objects while the activity in their prefrontal lobes was monitored using a brain scanning techniques called near-infrared optical spectroscopy. The brain scans showed that all groups used both brain hemispheres for creative tasks, but that the activation of the right hemispheres of the schizotypes was dramatically greater than that of the schizophrenic and average subjects.

Further research has added to the evidence: Swiss neuroscientist, Peter Brugger, has shown that everyday associations, such as recognizing the car key on your keychain, and verbal abilities, are associated with left hemisphere function while novel associations, such as finding a new use for an object or navigating a new place, are associated with the right hemisphere. A new study too, led by John Kounios of Psychology at Drexel University, has revealed distinct patterns of brain activity, even at rest, in people who tend to solve problems with a sudden creative insight compared to people who tend to solve problems more methodically. The creative solvers exhibited greater activity in several regions of the right hemisphere and this activity occurred even during the preliminary relaxed time before the experimental task began. Kournios has concluded that, 'problem solving, whether creative or methodical, doesn't begin from scratch when a person starts to work on a problem. His or her pre-existing brain-state biases a person to using a creative or methodical strategy.' It really does seem then that some people, like Einstein and Helmholtz, are 'right-brain types' and that their (and our) ability to access creative inspiration depends on a background level of right hemisphere activity.

SCHIZOPHRENIA ~ψ~

Schizophrenics often share the visual difficulties displayed by patients without a right hemisphere. In fact it is so common that a 'Draw-A-Person Test' is routinely conducted to diagnose schizophrenia. When asked to draw a representation of a figure, those suffering from the illness will leave out obvious anatomical parts like the hands or eyes and frequently do not distinguish between the sexes. It cannot be coincidental that such similar symptoms exist in schizophrenics and those suffering right hemisphere damage.

Perhaps in schizophrenics a greater split in function between the hemispheres leaves the dominant left hemisphere more isolated from the influence of the right. Less hemispheric co-operation, coupled with left dominance, results in greater dysfunction. Many causes have been suggested for schizophrenia including changes in brain biochemistry, but this may turn out to be something to do with the differences in chemical response between the two hemispheres.

Though schizophrenia is complex, certain symptoms are typically present in those suffering from the disease. These include hearing voices, deterioration in consciousness and the breaking down of the sense of 'I'. Auditory hallucinations are the most prevalent symptoms. Patients may hear one or many voices, and they are often identified as Gods or devils. These voices, whoever they are ascribed to, are sometimes regarded as benevolent and may help the individual and can even be ecstatically enjoyed. More often however the voices appear to persecute and they may become greatly feared.

One of the most interesting aspects of the voice phenomenon is that, while it is not even slightly under the control of the individual's normal sense of self, the character and subject of the voice is dependent on social environment. For instance, on the West Indian Island of Tortola, children are brought up to believe that God controls each detail of their life. When members from this society suffer mental illness, they frequently describe experiences that relate to this religious conditioning. They may hear loud prayers, hymns and commands from God ringing in their heads and even have feelings of being burnt in hell. Thus the detail and tenor of the symptoms does depend on past and external associations yet the manifestation appears to come from another place within that is not associated in any way with past experience.

The auditory hallucinations can begin as thoughts that then transform themselves into vague whispers, which then gradually become louder and more dominant. Sometimes outside noises such as the wind or rain transform themselves into patterns of words, and sometimes sufferers can actually feel their thoughts dividing. These latter reports are truly intriguing; it seems that the individual is observing a separation of function or at least becoming aware of the other half of their twin self whilst their normal self is still present. (In ancient Greek, the word for insanity was 'paranoia' which had nothing to do with present associations of persecution but literally meant having another mind alongside one's own.)

The voices, as we have noted before, not only have a dramatic authority and a feeling of religious profundity, but also appear to function at a greater speed than the familiar left or dual hemisphere system. Patients complain of having their usual thought streams anticipated by their voices. Some say they never get the chance to think for themselves – it is always done for them and when they try to read, the voices read to them first. When speaking too, they commonly hear the thoughts spoken in advance. It appears then that our second system, when freed from suppression, operates in a faster way than our normal system. Furthermore, there are indications, as we have seen, that liberated right hemisphere processing facilitates access to much greater memory. In times of severe emotional and physical trauma there can be almost total recall of our life's events. Near death experiences suggest that this occurs as part of the dying process. It may be that the left hemisphere finally gives up at these times of crisis allowing the right to take over.

Schizophrenics commonly experience a breakdown of their sense of 'I'. They can become confused in conversation, unable to work out whether they or others are talking or indeed how much of themselves is in themselves or in others. Patients describe how they need to sit still for hours at a time just to find their thoughts, and report the strenuous effort it takes to find their own ego identity for even a few brief moments. This can cause tremendous anxiety but what is it that is actually anxious? Is this great struggle the left hemisphere 'sense of self' desperately attempting to regain control? Its loss elicits much fear, and the limbo 'no man's land' of partial control is, of course, one of confusion.

Often, along with the loss of 'I', there is disorientation in respect of time. Patients may complain of time slowing down, being suspended or stopping altogether. For one individual, the spatial sequence of day and night 'had no shape in my memory'. Time consciousness is, of course, a very linear function. The problems schizophrenics have with time indicate that it is somehow allied to left hemisphere processing, and we can confirm this from our own experiences. Stepping out of time is not limited to schizophrenics. Time can stand still when we become totally immersed in something – when for instance the landscape becomes so stunning that our thinking process ceases. And there are parallel reports from the shamanic world. Carlos Casteneda calls the process 'stopping the world'. In 'Journey to Ixtlan' he writes of a time in which he:

'[...] had no thoughts or feelings. Everything had been turned off and I was floating freely. I stayed on that hilltop in a state of ecstasy for what appeared to be an endless time, yet the whole event may have lasted only a few minutes, perhaps only as long as the sun shone before it reached the horizon, yet to me it seemed an endless time. Never in my life had I such a divine euphoria, such peace, such an encompassing grasp, and yet I could not put the discovered secret into words, or even into thoughts, but my body knew it.'

The inability to put the experience into words, together with a stepping out of the time dimension, indicates a crossing from left to right dominance. Is 'stepping out of time' then merely a perceptual anomaly initiated when something within consciousness shifts, or is our perception of time something to do with left hemisphere dominance? What happens to time when we access right side function?

Subjects coming out of hypnosis are often surprised how much time has elapsed. In the hypnotic state (when left hemisphere influence is reduced) time stands still. Time may thus be nothing more than a linear ordering by a dysfunctional part of our neural equipment. Could it be that time, as many have theorised, is illusory? Logically we would say no, of course it's not. But it is our left hemisphere self that is taking this position. Our right hemisphere self, which may have a truer perspective on this matter, has a very different perception of time. This is an extremely interesting area that deserves thorough experimental investigation.

One other common feature of schizophrenia has been termed body image boundary disturbance – a disintegration of body sense. One patient described the following sensations:

'When I am melting I have no hands, I go into a doorway in order not to be trampled on. Everything is flying away from me. In the doorway I can gather together the pieces of my body. Why do I divide myself in different pieces? I feel I am without poise, that my personality is melting and that my ego disappears and that I do not exist any more. Everything pulls me apart.'

This experience is very reminiscent of states one is able to access in deep meditation (a mind tool that accesses right hemisphere function). Body image can collapse and the shock of this is often followed by an immediate fear that can abruptly halt any progress. This fear is like a barrier that if passed though can lead on to sublime states of bliss. Such experiences are consistent with the hypothesis that it is the left hemisphere 'sense of self' that is fearful and left hemisphere control that is limiting. Again such phenomena point to the existence of a second system within us – a right hemisphere self that not only has a different sense of perception but also, it seems, a different level of physiological function too.

Some schizophrenics are capable of tremendous feats of endurance. They can work endlessly and need little sleep. Catatonics may maintain awkward positions for days – positions that we couldn't hold for minutes. Julian Jaynes suggests that these factors indicate that fatigue is largely a product of the subjective conscious mind. And this idea is compatible with our findings that the left hemisphere requires more sleep than the right hemisphere.

Schizophrenics are also more alert to visual stimuli and can take in much more detail – in fact they are often drowning in data. They see every tree and every leaf. This may confuse their overall perception of the forest, but they also seem to have a more immediate and absolute involvement with their physical environment. Jaynes calls this a greater 'in-the-world-ness'. Is this a level of connection that us 'normal' folk so fundamentally lack and are subconsciously striving for? There is certainly a link here between the visual sensibilities of schizophrenics and autistic savant artists like Stephen Wiltshire who can hold incredibly detailed pictures in their minds and reproduce them later on paper.

Some experimental evidence adds weight to the idea that a breakdown in the balance between the two hemispheres is involved in schizophrenia. Electroencephalogram (EEG) studies have shown that over an extended period, normal individuals have slightly greater activity in the dominant left hemisphere. The reverse is the case in schizophrenics. They show more activity in the right hemisphere. EEG studies have also shown that normally our brain activity switches between the two hemispheres approximately every minute. In schizophrenics this switching occurs about every four minutes – a huge difference. This may indicate that schizophrenics 'get stuck' in one hemisphere and thus cannot shift between modes of processing as fast as the rest of us. Another avenue of research, that analysed skin conductivity in mentally depressed patients and schizophrenics (John Gruzelier, 1976); also supported the case for lateralised dysfunction being a major player in these conditions. Gruzelier's work in particular implicated left hemisphere problems in schizophrenics.

There are many theories of the whys, wherefores and hows of schizophrenia. Some have implicated a stressful environment and childhood trauma, some heredity and yet others abnormal biochemistry. All these factors may have a role in the illness but perhaps the common denominator is they all could contribute to a disturbance in the balance between right and left hemisphere activity. Extensive inhibition of the left temporal cortex, whether disease, genetic inheritance or stress causes it, can release the right brain from the normal level of suppression.

This conclusion is supported by highly significant research into epilepsy. When a lesion on the left temporal lobe causes epilepsy, 90% of patients also develop paranoid schizophrenia often with massive auditory hallucinations. When temporal lobe epilepsy is caused by lesion damage on the right side, less than 10% develop such symptoms.

The partial release of right hemisphere suppression evidently leads to confused and dysfunctional states. Suppressed elements of the right hemisphere self leak through but, along with the voices that play havoc with schizophrenics, there are, as we have seen, functions that appear to be more efficient than normal. These enhanced abilities are not only associated with schizophrenics.

MULTIPLE PERSONALITY SYNDROME ~ψ~

The emergence of enhanced abilities in some cases of schizophrenia, autism and physical left hemisphere damage suggest that we all have a latent second level of functioning. Within the extraordinary phenomenon of multiple personality disorder there are further anomalies that strengthen this view.

Although the first cases of multiple personality disorder (MPD) were studied back in the 19th century, the disease has only recently been fully accepted by the medical establishment. In 1985 the American Diagnostic and Statistical Manual of Mental Disorders acknowledged its existence and defined it as:

The existence within a person of two or more distinct personalities or personality states (each with its own enduring pattern of perceiving, relating to, and thinking about the environment and self). And furthermore, at least two of these personality states recurrently take full control of the person's behaviour.

The interest in the syndrome has expanded in response to the pioneering work of Roger Sperry, who in 1980 was given a Nobel prize for identifying specific functions of the brain that were lateralised either to the left or right hemispheres. This research was so far reaching that he drew the conclusion that human beings were really two people in the same body. He said that we were:

'two separate spheres of conscious awareness, that is two separate conscious entities running in parallel in the same bony cranium, each with its own sensations, perceptions, cognitive processes, learning experiences, memories and so on'.

The reader will note the close correspondence between Sperry's conclusions and the definition of MPD.

This idea that within us there are in fact two processing systems is not new. Back in 1844, Dr Arthur Wigan published a book entitled 'The Duality of the Mind' in which he postulated that the two cerebral hemispheres were capable of separate volitions that were often in conflict. He concluded, like Sperry a hundred and fifty years later, that the two hemispheres were separate brains, one subordinate to the other, but nevertheless perfectly capable of acting independently. Even the Greeks, back in the Classical Era, had their concepts of dualism, two minds and the other self.

All this seems relatively benign but those individuals suffering from MPD are subjected to much confusion. Imagine being surrounded by friends that belong to the other self and having no knowledge of who they are. Or being congratulated for aspects of your life (like piano playing) that you do not know anything about. Or being accused of lying or even arrested for crimes you don't realise you have committed. It really is a Jeckll and Hyde scenario. There are many blank spaces in a multiple's memory.

It even appears that the other self can emerge at night and can go off walking. A hostess of a tavern near Basel, after accusing her employees of stealing the takings, found it was her second sleepwalking self that was the criminal. She found broken glass and blood on her bedclothes and the money hidden away in the roof, but her original self remembered nothing of the nocturnal activity. In a similar case, a young woman became an arsonist in her somnambulist state. She set fire to rooms in her own house and those nearby and was totally unaware of this hidden side of her life. Even when she was caught, her waking self had no recollection of these events.

Like many of these odd ailments that are afflicting modern man, MPD is becoming more common. Not only is the incidence of cases increasing, but also the number of personalities contained within each individual seems to be multiplying. Clinicians report an average of eight to thirteen personalities and in a few cases of 'super multiples' over a hundred personalities have been uncovered in one individual. Why is this happening now? Is this part of the bigger picture? Is it another example of degeneration, and can it be linked to the huge chemical experiment that modern life imposes, or is it just a response to the growth in the number of psychotherapists. The truth is probably a mixture of the two.

If increasing degeneration of the left hemisphere is implicated in MPD, it may explain some of the anomalies that crop up in association with the syndrome. Dr Caul, the MD of Billy Milligan, a multiple who was the first person judged not guilty of a major crime on account of his MPD, noted that multiples are exceptionally perceptive. *"They can smell a liar at a thousand paces in one ten-thousandth of a second"* he said. And other talents have emerged too, including way above average visual and auditory memory, fine artistic ability, and the baffling tendency to heal faster than normal. What is even more extraordinary is that only some of the personalities within an individual may possess these talents.

There are cases in which just one of an individual's selves will possess, for example, the ability for extra-sensory perception. There are also cases in which it appears that the individual can use the different personalities within for multi-tasking. 'Cassandra' was a multiple who happened to be studying brain sciences. She reports:

'When I am writing a paper on dichotic hearing, one of the others is composing the proposal for 'my' master's thesis. Someone else has prepared dinner and will later clean up the kitchen while I sleep. [...] We share the body so the time I am at the typewriter limits the others' use of the body. It does not prevent any one of them from using the brain to plan, design or compose'.

This case is perhaps reminiscent of the multi-channel abilities of Julius Caesar noted in the first chapter. It seems that both Cassandra and the famous emperor had found a way to access the vast capacity of the brain that the rest of us rarely access.

The different personalities within an individual 'multiple' are usually distinct with little cross over between them. Billy Milligan, for example, only knew one language, English, but his alter ego known as 'Regan' was fluent in Serbo-Croat, and his 'Arthur' knew both Swahili and Arabic. Some personalities within one individual show physical differences like reversed handedness. It takes only a few seconds for a multiple to switch personalities and within that time a right-handed person can become left-handed. Studies of the brain wave patterns of multiples have revealed that the differences between their personalities vary as much as from one normal person to another.

Dr Robert Vito of Loyola University, by investigating at blood flow in different regions of the brain, found that the different personalities within a multiple show different patterns of biochemical balance. He has speculated that *the clinical switch from one to another may be the result of a chemical switch process involving the complex phenomenon of memory*'. It appears that each personality has its own biochemical setting that re-patterns the brain and alters body function. This not only affects handedness. Diseases and allergies can appear in one personality and not others. One doctor recorded changes in eye pressure and corneal curvature as personalities changed. Astigmatisms can come and go too and so different personalities within a single body can require different spectacles. One woman apparently suffered three menstrual periods per month – one supposedly for each of her alter egos and a multiple with diabetes lost the disease when her personalities changed. Epilepsy and dyslexia have been tied to distinct sub-personalities within a multiple, and those with MPD may age at a slower rate too.

The MPD phenomenon shows us that it is possible for something within an individual to switch over, resulting in fundamental body and brain changes. It appears that, despite the confusion it causes, multiples have greater access to right hemisphere functioning. If this functioning is more efficient it may explain how multiples can tap into enhanced abilities such as better eyesight, healing more quickly and why diseases do not impinge so readily. We predict that people with MPD have lower internal steroid activity. This would cause the thymus to be more active conferring a greater level of immune function. As the chemicals produced in the pineal gland to a large degree, control steroid activity, we expect multiples to have above average levels of pineal activity too.

(Under our normal regime, left hemisphere control may be, in effect, inhibiting the pineal and reducing the function of the thymus. We know the thymus gland becomes smaller and less functional after puberty and that it can re-grow when the steroid levels in the body are reduced, but why should this be? As there seems to be no evolutionary advantage for a suppressed immune system, such a dysfunctional arrangement seems to be another anomaly brought about by the dominance of an only partially functioning left hemisphere.)

We would all like to have access to enhanced abilities (extraordinary memory, enhanced visual acuity, the ability to multi-task and to heal more quickly) without suffering the negative effects of MPD. And perhaps it is possible to tap into different levels of brain function. One 31 year old women, according to a case history presented by Gott, Hughes and Whipple (1984), had the ability to consciously switch between two different emotional patterns or personalities. The first was a logical, mathematical and organised businesswoman who was upset by inefficiency while the second was relaxed, sexy and enjoyed sports. The traits of the first personality are typical of left hemisphere activity while the second are typical of the right. Sure enough, electroencephalograms confirmed that there was more activity in the left hemisphere when the women was in her business women persona and more activity in the right when she became the sexy one. The researchers concluded that this woman could work in her left hemisphere persona or her right whenever she willed it.

The internal liar

One question that has haunted psychiatrists and philosophers alike is the issue of German concentration camp workers. How could 'normal' inhabitants of villages around the camps have taken on jobs within the camps, be subjected to and involved in all the horrors there and then gone home to their families at the end of the day as ordinary fathers? It seems natural to conclude that these workers were deranged but, except for a few extremists, all the evidence suggests that the majority of these camp workers fell within what is regarded as a normal mental category. David Pedersen, author of 'Cameral Analysis', argues that a possible explanation could lie in the ability of these workers to separate their personalities. He deduces that:

'if this is so, then it would appear that their left logical hemisphere has been persuaded, or manipulated, into believing a certain belief or behaviour was logically correct. Once this is established in the mind of the individual, then it appears possible to carry out any form of atrocity under the cloak of it being justifiable'.

History is full of such bizarre double standards and it is particularly prevalent when religious fundamentalism is combined with intolerance. The results are usually horrible – witch burnings, fatwahs, inquisitions, excommunications, genocide etc. And it is all done in the name of goodness. It really does seem that something is inherently sick within our consciousness system.

There was some very interesting research work carried out by Sperry, Gazzaniga and Le Doux in the late 1970s that sheds light on this conundrum. These researchers conducted tests on patients that had their two hemispheres isolated by severance of the corpus callosum (a successful cure for epileptic seizures). These individuals became 'split-brained' – their two sides worked independently. Different images were presented to these patients, side by side, on a screen, and being split-brained, their two sides responded differently.

In one trial, a spoon was projected on the left side of the screen so that only the patient's right hemisphere perceived the image. When asked what he saw he replied "nothing" but at the same time his left hand (controlled by the right hemisphere) picked out a spoon from an assortment of objects beneath the screen. This shows that, though the right hemisphere was functioning properly, the left hemisphere conscious self was ignoring the true perceptions and actions. In contrast, when the name of an object was flashed on the right side of the screen (so the perception went to the left hemisphere) the patient could describe the object and pick it out correctly.

When different objects were projected on the left and right sides of the screen at the same time, the patient picked out what he saw on the left with his left hand and what he saw on the right with his right hand, without being aware of any conflict. (The left image is perceived by the right hemisphere, which controls the left hand and vice versa). When the patient was then asked why his left hand had picked out an object that differed from the choice of his right, and his verbal affirmation of that choice, he replied that he must have done it "unconsciously".

This is of major interest for it appears that the left hemisphere processing system is inventing reasons to explain away inconsistencies in right/left perception.

A further trial run by Gazzaniga and Le Doux was even more enlightening. The patient 'P.S.' was given two different tasks at the same time – one to each hemisphere. On a screen, a snow scene was presented to the right hemisphere, and a chicken claw to the left. Beneath the screen was a series of pictures from which the patient had to pick out the ones most appropriate to the screen images. P.S. responded correctly by picking out the picture of a chicken with his right hand. The left hemisphere had perceived the chicken claw and instructed the right hand to point to the picture of the chicken. At the same time his left hand pointed to a picture of a shovel because his right hemisphere associated a shovel with the snow scene. All well and good but, when asked what he saw, he replied that he had seen a chicken claw and picked the chicken picture but added that you have to clean out the chicken shed with a shovel. *Conclusion – even though the left hemisphere was not consciously aware of the right hemisphere's view of the snow scene, it did incorporate the right hemisphere's response (pointing to the shovel) into its explanation, even though it had no idea of why it was doing this.*

Gazzaniga and Le Doux elicited this type of response in many of their experiments. The left hemisphere identified why and what it had picked out, but then seamlessly incorporated the right hemisphere's response into its story. It did this in a very matter of fact way. It did not offer its suggestions in a guessing vein but rather as a statement of fact – the left hemisphere, in effect, *was making up stories and believing them to be correct.* Of course, this trial was conducted with split-brained patients and we must be cautious about extrapolating the results to the rest of humanity, but similar results have been obtained too, in normal, non split-brained patients undergoing hypnotherapy.

A patient was asked, whilst under hypnosis, to open an umbrella lying on a nearby table after she came out of the trance state. This she did and when asked why she opened the umbrella, she made up a story in just the same way as Gazzaniga's and Le Doux's patients. David Pedersen has deduced that during hypnosis the left hemisphere is being inhibited so that consciousness is limited to the right hemisphere. In split-brained patients connections between the left and right brain were severed. In both these examples therefore, the left brain is exposed, and we can clearly see what it is up to. It, in effect, lies without even realising it. Are we subject to this sort of inconsistency all the time? If so, how would we know? We can sometimes see the mental blockages and blind spots in others but rarely in ourselves.

HYPNOSIS ~ψ~

David Pedersen recalls another hypnosis experiment that was conducted on a BBC television programme. Professor F. Frankel of Harvard Medical School hypnotised a female volunteer and asked her to go back in time to a happy event in her childhood but, instead of being joyful, the subject started to cry. Under hypnosis, she returned to the events surrounding the birth of a sibling, which she had been told repeatedly, was going to be a very happy time. Her 'left hemisphere verbal processing system' had registered the event as a happy one but what actually occurred on the day was that she had been frightened by her mother developing a severe pain. The mother was then whisked off in an ambulance leaving her alone and terrified until her father returned from work. She thus had two realities of the same event. The real experience and the verbal construct made up of other people's expectations and influence.

Remember Sperry's words: '*two separate spheres of conscious awareness*'. Do we all have two realities? A right hemisphere reality based on true perception and experience, and a left hemisphere reality based on verbal stories it has made up to cover up the conflict, and in which it believes totally. If this is correct, the implications are extremely consequential. Not only does it explain the conundrum of the German concentration camp workers but it also explains other anomalies like why 98% of soldiers, despite being trained to kill, will shoot wide of the mark when confronted by an actual human enemy. Again this indicates that the left hemisphere is not functioning like a good hemisphere should. We have a controlling side that can be logically persuaded to kill and much more, but beneath this is a hidden side that has an entirely different perspective.

After assembling a mass of evidence, some of which we have reproduced here, David Pedersen states:

'Because a person in an hypnotic state, particularly a deep somnambulistic state, exhibits the same cognitive and behavioural functions now attributed by neurological research to the right hemisphere, then it would suggest that the state of hypnosis is right hemisphere-orientated. We can also postulate that the mechanism of going into an hypnotic state involves either a shift into right hemisphere function, or an inhibition of the left hemisphere, or both.'

He also points out that the variation of depth of the trance states that are entered could be explained by the degree of inhibition of the left hemisphere. *'A light trance would involve only a partial inhibition, whereas the extreme somnambule depth would correspond to a much greater degree of inhibition'.*

If our hypothesis about right hemisphere latent function were correct, we would expect some unusual abilities to emerge when the left hemisphere is inhibited by hypnosis. And it does. Richard Bandler, the co-founder of neuro-linguistic programming, is also one of the world's leading authorities on hypnosis and its applications. He found, in experimental hypnosis trials, that people in a trance state could do things that they couldn't do in their normal state of consciousness. For example, they could carry out phenomenal computations with numbers (just like autistic savants), and those with visual impairments could see well without their glasses. Bandler also invented 'hypnotic telescopes' – a hypnotic device that enabled subjects to read at greatly increased distances. These abilities are proven, demonstrable, and, of course, work with the same brain that doesn't appear to have these enhanced functions in the non-trance states. So what is the difference? What is being accessed? If Pedersen is correct and the trance-state not only inhibits the left hemisphere but also allows uninhibited right hemisphere function, the answer would seem to be obvious.

Another enhanced function that is released by hypnosis is dramatic memory recall. Childhood events, such as the presents one received at Christmas, can be retrieved in stunning detail. The emotional content and feelings of those distant times can be recalled too. A.R.

Luria and E.G. Simernitskaya of Moscow University (1977) have shown that memory consists of two separate cognitive functions. The first is intentional memory – the type that we use consciously when, for example, we revise for an exam. The second is involuntary memory in which events, and all the phantasmagoria of life, are absorbed without any conscious effort. Luria and Simernitskaya believe that intentional memory is a left hemisphere function while involuntary memory belongs to the right.

Suppression of the left hemisphere by hypnosis and, as we noted earlier, by restricting sleep allows greater access to our deep memory store. Most workers in this field would conclude from this data that both hemispheres have their specific functions and both are working 'as they should'. We question this assumption. We believe that so-called intentional memory is all that the left can manage. The right, in contrast, has a much greater facility. In fact all our past experience may be buried somewhere within its arena but our access to it is hampered by the dominance of the left.

There is evidence too that the right hemisphere self has a greater facility with visual data. Visual dreamlike images and hallucinations are easily accessed within the hypnotic state and indeed dreaming itself is regarded as a right hemisphere function. We know that damage to specific areas of the right hemisphere stops dreaming in those who previously dreamed normally, and such individuals also lose the ability to visualise whilst awake. Experiments, in which electrical currents were applied to these same areas of the right hemisphere, elicited visual illusions and memory flashbacks. And further experiments stimulated double consciousness – in one, a patient was aware of being in an operating theatre talking to a doctor, at the same time as they appeared in a dream scene. The electrical current stimulated the right hemisphere to such an extent that normal left hemisphere suppression was overridden. (Wilder Penfield, 1959) As stimulation of the corresponding areas of the left hemisphere only affected speech and muscle movements, it seems that this research too indicates that the right hemisphere possesses more complex functions than the left and that these are normally held in check by the left's inhibitory control.

The inhibiting effect of the left hemisphere is illustrated clearly when we awaken from sleep and promptly forget our dreams. We may have a fleeting sense of dreaming something, but when the left brain clicks in, the dream images are largely lost. The left on waking re-establishes dominance. The dreams are lost because the dreams did not happen there. They happened in right brain consciousness. As we all know, it is quite tricky re-accessing our dream memories. The harder we try the less we remember – the more we engage our rational mind the further we move from the place of the dream experience. Yet the right hemisphere is still a part of us. The dreams are in there somewhere and with practice we can improve our recall by such techniques as engaging our visual memory and not thinking.

Interestingly, research on split brain patients has indicated that, despite a paucity of dreams, fantasies and symbols, the left hemisphere does have some dreaming capacity. Psychoanalyst Klaus Hoppe analysed the dreams of twelve patients and found that the content was significantly different from normal dreams. They were utilitarian, unimaginative and tied to reality. This reinforces the view that the left hemisphere is a very linear, unelaborate processor in comparison to the right that is capable of much greater visual richness and much else besides.

People vary in their susceptibility to hypnosis. Studies have shown that art students are easier to hypnotise than science students. This is consistent with the premise that, in general, scientists, being more predisposed to working in logical, left hemisphere mode may find more difficulty in shifting into their right side. Older people tend to be harder to hypnotise too. As we age we become more 'stuck in our ways', which is another way of saying stuck more in our left hemisphere mode of processing. This is to be expected after years of right hemisphere suppression. As we age there is a general tendency to talk more, to become less imaginative and more fearful (all indicative of increased left hemisphere dominance). In contrast, children tend to be highly imaginative and reckless. They live much more in a fantasy world of imagination – a right hemisphere mode of operation that is much richer and more complex than the linear, step by step mode. We know that handedness and thus the establishment of left hemisphere dominance increases with age so it is reasonable to conclude that children for the most part have more access to right hemisphere function than adults. And this is reflected in children being particularly easy subjects to hypnotise.

The problem with words

The right hemisphere mode of function, as we have noted, has a greater facility in working with images and pictures – a much more vivid, faster and direct mode than stringing a description together with words. Although in today's world we regard words and speech so highly there are some real problems with word communication. We never know whether the meaning and feelings associated with a word exactly corresponds to another's interpretation. Richard Bandler and John Grinder (both originators of N.L.P.) have even said that:

'it would be easier to do therapy in a foreign language! That way you would not have the illusion that the words you heard had the same meaning for the person who uttered them as they have for you. And believe me it is an illusion.'

In their experience they have found that most learning and change takes place at an unconscious level. And when they refer to the unconscious level they are really, in our view, talking about right hemisphere function. They even recommend that their clients '*make use of the natural processes of sleep and dreaming to review any experiences that might have occurred*' in their workshops.

Bandler and Grindler use direct connections to the unconscious mind (the right hemisphere) to circumvent the problem with word communication. This is particularly appropriate when treating relationship problems because, as they say, '*couples usually get into trouble with words, because people are not very good with words*'. Perhaps it is just that words are very divorced from our real internal experience. There is something like a descending staircase of interpretation involved here – on the top step there is the unsullied, actual experience that is, on the next step down, interpreted in terms of emotional response. On the step below lie our feelings and images about the event and only on the bottom rung are the words. Words are a long way from the actual experience. They are a very poor approximation – no wonder we get caught up in knots. It can be a great relief, too, when we leave behind the whole word system for a while. We re-energise in silence. This is particularly noticeable on silent meditation retreats when word use is restricted or abandoned altogether for periods of days or weeks. Much joy manifests when we are released from the restrictions of the left hemisphere mode of processing. Why should this be so, if the left hemisphere were functioning properly? Fritz Perls, the Gestalt therapy man, evidently referring to the left hemisphere word processor, even said: '*Lose your mind and come to your senses*'.

Our left hemispheres, to put it bluntly, are not only liars but also a little stupid. This is quite clearly demonstrated by a hypnotherapy session that is included in Bandler and Grinder's book 'Frogs into Princes'. We will quote it at length because the issues it raises are of such importance.

'*A man came in once and said there were all kinds of things that stood in the way of him being happy. I said, "Would you like to tell me what those things are?" And he said, "No, I want to go into a trance and change it all, and that is why I came for hypnosis." So accepting all behaviour, I did an induction, put him into a deep trance, sent his conscious mind away, and said, "I want to speak privately with your unconscious mind." I have no idea what that means. However when you tell them to, people do it. They talk to you and it's not the one you were talking to before, because it knows things the other one doesn't know. Whether I created that division, or whether it was there already, I have no idea. I asked for it, and I got it.*

In this particular case, his conscious mind was, to put it as nicely as I can, inane. His unconscious resources, however, were incredibly intelligent. So I said, "What I want to know from you, since you know much more about him than I do, is what change is it that he needs to make in his behaviour?"

The response I got was, "He's a homosexual".

"What change does he need to make?"

"He needs to change it, because it is all based on a mistake."

"What mistake?"

The explanation that I got from his unconscious mind was the following: The first time he has ever asserted himself physically, in terms of trying to defend himself against violence, was when he was five years old in hospital to have his tonsils out. Someone put the ether mask on his face, and he tried to push it away and fight back as he went under the anaesthetic. Anaesthesia became anchored to the feeling of being angry. After that, every time he began to feel angry or frightened and started to strike out, his body went limp. As a result of this, his conscious mind decided he was a homosexual. He had lived as a homosexual for about twenty-five years.

His unconscious resources said, "You must not let his conscious mind know about this mistake, because knowing that would destroy him". And I agreed with that. The only important thing was that he makes a change, because he wanted to get married. But he couldn't marry a women because he knew that he was a homosexual. His unconscious mind would not allow him in any way to become conscious of the fact that he had made this mistake, because it would have made his whole life a mistake and that knowledge would have utterly destroyed him. It wanted him to have the illusion that he grew out of it and grew into a new behaviour.'

Bandler arranged, with this subject's unconscious mind during hypnosis, to have his outer self make the changes that he needed to make, but only after a 'spiritual, drug, cosmic experience' occurred in his outer life. This gave his conscious mind the platform to make the change without realising the fundamental fault lines that his hypnotherapy session uncovered. But why should the left hemisphere need such subterfuge and mollycoddling?

We can see clearly from this example, that the left hemisphere self makes up stories that are not based on the truth of experience. It does its best but its best is not very good and being dominant, our whole lives can be based on its lies. It suppresses and appears disconnected from a truer perspective provided by right hemisphere function. Is it any wonder therefore that our interpersonal relationships are fraught with difficulties and society as a whole is under increasing strain? So many dysfunctional but dominant left hemisphere selves interacting in increasingly crowded conditions!

CONFABULATION ~ψ~

In a study of normal healthy individuals, Jerome S. Bruner and Leo Postman (On the Perception of Incongruity: A Paradigm, 1949) demonstrated a quite extraordinary level of self-deception.

In the experiment, twenty-eight students at Harvard and Radcliffe were randomly shown five different playing cards and asked to identify them. Some of these cards were incongruous – their colour and suit were reversed, for instance, the five of clubs was red. The results were intriguing. There were four kinds of reactions to the rapidly presented incongruities. The first was a 'perceptual denial' of the incongruous elements. Faced with a red six of spades, for example, subjects reported with considerable assurance, 'the six of spades' or the 'six of hearts', depending upon whether they were colour or form bound. In both instances the perceptual result conforms to past expectations about the 'normal' nature of playing cards. A second way of dealing with the incongruous stimuli was a colour compromise. For example, a red six of spades was reported as 'a purple six of hearts', and in another instance, a red six of clubs was seen as 'a six of clubs illuminated by red light'. The third type of reaction was even more bizarre. The incongruity so baffled subjects that their perceptual ability broke down almost entirely. One frustrated subject said: 'I don't know what the hell it is now, not even for sure whether it's a playing card'. The fourth reaction was an actual recognition of incongruity – the odd card was correctly identified.

It most cases then it turns out that perceptual organisation is powerfully determined by expectations built from past experience. When such expectations are violated, the perceiver resists recognition of the unexpected or incongruous. The perceiver in effect finds it hard to come to terms with the new reality and tends to make up stories to cover over the incongruity.

Scientists have now applied the term 'confabulation' to this phenomenon and much more has been done to uncover the extent of the problem. In a one revealing experiment, Timothy Wilson of the University of Virginia and his colleague Richard Nisbett set out four identical garments and asked people to pick which they thought was the best quality. If there is no other reason for making a choice, people will usually prefer the rightmost object in a sequence; and sure enough four out of five of the participants did select the clothes on the right. Yet, when asked why they made their choice no-one mentioned position. Instead they said the colour, texture or fineness of the weave was superior. Could it be then that, though we routinely make our decisions subconsciously, our rationalisation of our choices might be pure fiction?

Further work by Lars Hall of Lund University in Sweden adds to the evidence that this remarkable scenario is indeed the case. In his experiment participants were shown pairs of cards displaying faces and asked to choose the most attractive. However, by sleight of hand, the chosen card was swapped for the rejected one. This subterfuge often went unnoticed and, when asked why they had made their choice, subjects came up with many elaborate explanations involving hair colour, the look in the eyes and the assumed personality of the substituted face.

In a recent paper (The Evolutionary Biology of Self-Deception, Laughter, Dreaming and Depression: Some Clues from Anosognosia), the well-known neurologist Vilayanur Ramachandran makes the connection between such confabulation and the left hemisphere. He has confirmed that, under certain conditions, the left side of the brain is not only unable to update its reality but also that it unconsciously lies to maintain its subjective reality however untenable the objective evidence may be.

He cites examples from cases where patients have suffered right hemisphere stroke resulting in a paralysis of the left side of the body. This damage reduces the influence of the right hemisphere, effectively amplifying the dominance of the left. These patients can suffer from a condition called anosognosia, in which they are unaware of the reality of their condition and maintain an extraordinary level of self-deception about their incapacity. For example, when asked to perform an action with their paralyzed arm, they employ a whole arsenal of grossly exaggerated defense mechanisms to account for their failure (e.g. 'I have arthritis' or 'I don't feel like moving it right now'). In one case, Vilayanur Ramachandran conversed with a lady who has lost the use of her left arm:

VSR: Mrs D, can you clap? FD: Of course I can clap. VSR: Mrs D, will you clap for me? *(She proceeded to make clapping movements with her right hand as if clapping with an imaginary hand near the midline!)* VSR: Are you clapping? FD: Yes, I'm clapping.

The logic of those suffering from anosognosia can be surreal. It is not unknown for patients to deny ownership of an arm and yet, in the same breath, admit that it is attached to their shoulder. This is one of the most perplexing medical phenomena that one can encounter.

Anosognosia could possibly be explained as a reaction caused by a simple denial of something unpleasant, but this does not work for one simple reason: this phenomenon is rarely seen when the left hemisphere is damaged. Right-sided paralysis ought to be just as unpleasant for patients yet they rarely engage in denial. This asymmetry suggests that anosognosia is a *neurological* rather than a psychological syndrome. It is something to do with a left hemisphere operating without support from the right.

These experiments with anosognosia suggest that the correct knowledge of the patient's paralysis is being held somewhere in the brain but that access to this information is blocked. An ingenious experiment performed by an Italian neurologist Edwardo Bisiach on an individual suffering from neglect (an inability to perceive one side of their world) and anosognosia casts light on this. Bisiach took a syringe filled with ice-cold water and irrigated the patient's left ear canal. Within a few seconds, the patient's eyes started to move vigourously. Bisiach then asked the patient if she could use her arms. Surprisingly, the patient replied that she had no use of her left arm. The cold-water irrigation of the left ear brought about an admission of her paralysis. It acted like a truth serum.

Why did the water produce such a miraculous effect? One possibility is that the cold water 'arouses' the right hemisphere. (There are connections from the vestibular nerve projecting to the vestibular cortex in the right parietal lobe as well as in other parts of the right hemisphere.) Arousal of the right hemisphere makes the patient pay attention to the left side and to an arm lying lifeless and paralyzed.

Ramachandran has theorised that each of us has a tremendous need to impose consistency, coherence and continuity on our behaviour. We need a script – a thread of continuity in time. The left hemisphere is primarily responsible for imposing this consistency on to the storyline and this would correspond roughly to what Freud calls the ego. It's the left hemisphere's job to create a model and maintain it at all costs. If confronted with some new information that doesn't fit the model, it relies on defense mechanisms to deny, repress or confabulate; anything to preserve the status quo.

Ramachandran thinks the right hemisphere's strategy is fundamentally different. It detects anomalies – when the anomalous information reaches a certain threshold, the right hemisphere decides that it is time to force the left hemisphere to revise the entire model and start from scratch. If this is so, Ramachandran appears to be saying that the right hemisphere perceives reality whilst the left is lost in some conceptualised version of it.

In patients suffering from anosognosia, the left hemisphere carries on with its confabulation and denial, as it would routinely in a normal person. The difference is that these patients have lost the mechanism in the right hemisphere that would force them (when the stimulus became too strong) to generate a response to conflicting information. This leaves the patients lost in delusion, oblivious to their predicament and glibly explaining away *any* anomaly or discrepancy.

While this kind of behaviour is nakedly apparent in patients with damaged right hemispheres, we may all suffer from anosognosia to some degree. It seems to be a condition of the left hemisphere self. Indeed Lars Hall argues that it is a distinct possibility that 'people may confabulate all of the time'.

These startling observations and conclusions have been pushed on even further by the work of Allan Snyder. He has added to the evidence that implies that the left hemisphere does not experience reality but rather maintains an abstract version of it. While the right does deal with reality directly, its influence is greatly subjugated by the inhibiting effects of the left. As the left side of the brain is primarily responsible for our sense of self, this leaves us in a somewhat invidious position. It also calls in question whether cerebral dominance really is adaptive, as is presumed, or a symptom of a neurodegenerate condition. Of course if this is the case then it may well be difficult to convince the left hemisphere of such a novel scenario.

DEEPER INTO FEAR ~ψ~

There are subtle pathways within the brain/body that link consciousness to biochemistry. One investigation into the affects of mantra intonation, visualisation and deep meditation, by neuroscientist Ranjie Singh at the Saybrook Institute, Western Ontario Research Park, found that melatonin levels increased by up to 1000 % after a day of engaging in these practices. Thus something we do with consciousness can have a chemical effect, and chemical effects can lead to structural changes. These changes may be good; meditation can enhance wellbeing and wellbeing can enhance health, but the feedback loop can work the other way too. If there is a sense of self that is innately fearful, it will feed back into its own neural generator and change its biochemistry. This could potentially initiate a negative feedback loop; the greater the biochemical or structural damage, the more fear, which can in turn lead to more structural damage. Something like this may have been responsible for the decline in functioning of our left hemisphere, and this process would have contributed to hemispheric divergence too. In effect, our controlling 'sense of self' damaged its own generator as well as shutting out the other self that retained the potential for more complete function.

Like as a still millpond, a quiet consciousness system provides cohesion but if fear disturbs the calm, the ripples will inhibit full functioning. One of the sources, perhaps the main source, of the innate fear the left hemisphere feels is based on an anxiety about, or a suspicion of, the right hemisphere self. The dominant left now feels the other, the right, is something to be afraid of. While the brain was conditioned to deal with any dangers on the outside it was never equipped to deal with this internal glitch. It can attempt to cope with external challenges, such as tigers or tarantulas, but has no mechanisms for dealing with such internal

116

ones. The structures that are trying to deal with the problem are the very ones that are damaged and causing the problem.

There are clues to all this in the way our perceptions change as we grow up. It is quite common for children to have a sense of inner knowing, to feel a oneness with the earth and even have out-of-body experiences. Children who are particularly sensitive to such things are often ambidextrous – a faculty that is indicative of less cerebral dominance. Usually in these cases, cultural conditioning stifles these perceptions, and as the children get older access to such feelings and visions diminish. The decline of these 'wider' perceptions (that may be distorted, as they would still to a greater or lesser extent come through the filter of the left hemisphere self) almost certainly goes hand in hand with increasing left hemisphere dominance.

Child musical prodigies and other special children with outstanding mathematical and/or language talents (akin to autistic savants) can also lose a measure of their genius as they get older – and their left hemisphere control becomes more complete.

After puberty, consciousness changes to an adult mode and a more intellectual, 'stable', but fearful self increasingly dominates. This fear, for the most part, is kept below the surface but it manifests in all the things we are fearful about. It is projected onto all sorts of external things, particularly those that the media continuously feeds us via the news and television dramas. Most people do not acknowledge or see this basic fear. Fear to us is like water to fish. It is always present in the background, affecting our actions and decisions. We blithely go along vaguely thinking everything is somehow all right, or will become all right when we have achieved a certain goal (which is always changing).

The left hemisphere 'I' is particularly frightened of the emotional depth and connection of the right. It cannot really relate to or experience these things, and tries to block them. Emotions like love, if allowed through our inhibition system, can be sublime but the left hemisphere self can be frightened by even this. These fears of connection and of perceived external threats increase with age, particularly in males. It is always old men (with a lifetime of right hemisphere repression) that send young men, full of burgeoning testosterone, to war. History shows that it is much easier for these old statesmen to follow the pathway that leads to death and destruction than the much more emotionally risky route of trust, heartfelt communication and understanding. The sheer craziness of this has lead to something approaching 100 million people being slaughtered in twentieth century wars alone.

Colin Cafell, author of 'In Search of the Rainbow's End', has noted that denial of fear can become extreme. There are men who would rather get killed than admit, even to themselves, that they are scared. They would rather become violent and angry than risk someone seeing their tears. This fear pervades all society. In fact fortunes are made from it,

117

economies depend on it and industries survive by its perpetuation. Nearly everything we are exposed to in the media, newspapers, magazines, television, cinema, posters and radio, instills, to a greater or lesser degree, some level of fear in the citizen/consumer. The predominant focus of news reporting, often directly linked with prime time advertising, is on disaster, despair, hatred and betrayal. This makes us fearful of the world 'out there', but we are encouraged to compensate for this fear by buying something to make us feel good or safe. We are induced into shopping for a cornucopia of products and services by industries that tell us that we won't look good, be sexually attractive, be healthy, be part of the 'in crowd' or, with regard to the insurance industry, feel safe and secure in our old age without them. Fear and consumerism, the feel good factor and our economies are inextricable linked. And they are also linked to the internal glitch in our brain structure.

HIGHER FUNCTION ~ψ~

Fortunately not all is lost. Although it now seems clear that the left hemisphere on its own has limited abilities, we have seen from Sperry's work that we do have a second sphere of conscious awareness. We may not be able to access complete right hemispheric function at will, due to interference from the left, but it remains latently available, residing in a sort of dream world – unplugged, yet present.

Despite the problems with conscious access (it only touches our awareness in reveries and in dreaming), it is evident that right hemisphere processing does provide much of our complex function. The left on its own is only able to do comparatively simple things – its ability is linear and it operates on a one step at a time basis. We have seen that it does not even function well at speech and it is almost totally dysfunctional at other tasks, such as distinguishing males from females and coping with emotions. To engage in these more complex tasks therefore it has to 'borrow' function from the right. But there may be a price to pay for this. Complex 'right hemisphere function' filtering through a left hemisphere with limited capability may result in conflict, confusion and confabulation. For instance, what our hybrid sense of self recognises as emotions is perhaps the left's inadequate ability to work with right hemisphere function. The right can know something directly but this direct knowing may not match the left's linear analysis of the situation. This can cause conflict between the hemispheres that may panic the left into devising confabulated stories or, in the worst cases, cause neurosis and mental breakdown.

In many cases of autism, MPD and schizophrenia, sufficient disturbance of the left brain has allowed greater access to the more efficient right-self processing system, and this has released what are regarded as anomalous enhanced abilities. The tantalising question this throws up is whether it may be possible for all of us to access such phenomenal musical, artistic and perceptual talents without suffering the negative affects associated with these illnesses.

Allan Snyder's research is beginning to show that we can. It has been documented that autistic individuals can instantaneously compute the exact number of matchsticks that are tipped on the floor. To test this ability in normal people, Snyder inhibited the left anterior lobe of twelve participants with repetitive transcranial magnetic stimulation. (Degeneration of this part of the left hemisphere has been implicated in the savant condition.) Snyder found that ten participants improved their ability to guess the number of discrete items, and eight of these became worse at guessing when the magnetic pulses stopped. Snyder and his team think that the inhibition this part of the left hemisphere is allowing the brain's number estimator to act on raw sensory data without interference. Normally that part of the brain would turn the raw data (*reality*) into patterns and shapes. One of the team said: 'By inhibiting networks involved in concepts, we may facilitate conscious access to literal details, leading to savant-like skills'.

Other experiments carried out by Snyder's team have shown that inhibiting the left side with magnetic pulses can lead to improved numerical perception, mathematical ability and drawing skills, but these may just be the tip of an iceberg of increased function. Extra sensory perception, intuition, direct knowing and enhanced memory are all part of our human palette of experience. Because such talents are not universally accessible, these areas are controversial and difficult to study in laboratory conditions. However, scientists are beginning to catch up with these outer reaches of our consciousness systems too.

Take remote viewing for instance. Meticulous research by top flight American physicists Hal Puthoff and Russell Targ has established that we humans have the ability to view, or accurately imagine scenes, when given map references or similar cues. The CIA was sufficiently convinced to pour millions of dollars into secret remote viewing spy operations. It has been reported that, in the aftermath of the war in Afghanistan, the CIA even used remote viewers to try and establish the whereabouts of Osama Bin Laden.

Puthoff and Targ concluded from their experiments that we all have the ability to do this, but talented individuals find it easier to enter the framework of consciousness that allows them to see 'anywhere in the world'. The most important condition for success appears to be a relaxed, playful atmosphere and attitude. The worst thing a remote viewer can do is to interpret or analyse what he sees – a typical left hemisphere function. The sensory channel operating during remote viewing appears to make use of the unconscious and non-analytical parts of the brain.

Healing

Elizabeth Targ (a psychiatrist and daughter of Russell Targ) achieved parallel results in her profound study on 'distant healing'. Not only did she find that distant spiritual healing worked (it had a remarkable affect on extremely ill AIDS patients) but when she analysed which healers had the most success, she found it did not depend on beliefs or methodology. A Christian calling on Jesus was just as successful as a Lokota Sioux shaman, a Ch'i Gong master, a Jewish Kabbalist healer, a Buddhist and a New Age practitioner from the School of Healing Light. What they all seemed to have in common was the ability to leave their individual sense of self behind: Their ego sense of self was surrendered to some kind of healing force.

How this operates has still to be elucidated but it has been shown by Elmer Green (Copper Wall Project in Topeka, Kansas), that experienced healers have abnormally high electric field patterns during healing sessions. Patients undergoing 'hands on' healing sessions usually experience a warmth coming from the healer's hands. Lynne McTaggart, author of 'The Field', has argued that these phenomena may be physical evidence of the healer's greater overall 'coherence'. She believes healers are able to co-ordinate quantum energy and transfer it to the less organised recipient. To access this ability it seems necessary to leave behind the stream of disturbance caused by the chattering voice arising from the left hemisphere. She states that 'the left brain is the enemy of the field'.

The right hemisphere plays a major part in self-healing too. The visions, hallucinations and mystical experiences that sometimes accompany acute illnesses are indicative of right hemisphere activity. Historically, such experiences were actively encouraged as an aid to healing. The content of these visions may be distorted because the body/mind is *in extremis*, but that they occur at all is significant. At these times of crisis, it appears that the body can enter an alternate-healing mode that may be explained by less inhibited right hemisphere function. The immune system under right control is perhaps many times more efficient than the left-controlled system.

Once the disease is successfully combated, there can be a return to 'normality' – left dominance reasserts itself just as it does every morning when we awaken from our world of dreams. Often though, immediately after a fever (and the implied right hemisphere control) there is an afterglow; a residual feel-good factor that may be a pointer to the natural state of the right hemisphere self. This inability of the left hemisphere to cope and its tendency to give up in times of great stress or illness, may also have relevance to the state of calm many people reach shortly before dying. Similarly, during the First World War many of the millions of people exposed to terrifying situations went beyond their fear to experienced transcendent mind states. The tremendous stress appears to have caused the soldiers' left hemispheres to surrender its control mechanism, allowing the right's latent function to come through.

Intuition

There are parallels here with telepathy, intuition and insight. Again we find that the left hemisphere needs to be circumvented to fully access these function. Thought transference, for example, has been found to be most successful when participants are asleep, dreaming, involved in physical activity, or engaged in some relaxation technique such as meditation. In many cases the communications occur subconsciously, without the recipient (the left hemisphere self) being even aware of it.

There has been some significant work on our ability to intuitively solve problems. Gary Klein, an Ohio based consultant psychologist collected hundreds of stories in which rational and methodical pathways to decision making were bypassed in favour of intuitive hunches. These instantaneous decisions saved lives. In one instance Frank, a fireman in Pittsburg, was leading his team into a house to tackle a burning kitchen. Suddenly he had a 'bad feeling' and ordered his men out. Seconds later, the kitchen floor collapsed into a basement that Frank did not know existed. Intuition, it seems, can sometimes result in better decision making than logic and reason.

Jonathan Schooler, of the Learning and Development Research Center has corroborated this at Pittsburg University. He found that when experimental subjects were asked to make snap decisions on tasks ranging from choosing jam to selecting a university course, these decisions were more likely to be satisfactory and less likely to be reneged upon than when subjects deliberated beforehand.

That moments of insight can come to us instantaneously, and that decisions based on intuitive insight are more likely to lead to successful outcomes, is consistent with the view that there is another side to us that has more potential than we realise. It also highlights the two distinct modes of brain processing – one verbal, logical, linear and conscious, and the other non-verbal and below the level of normal conscious awareness. While some researchers are wary of assigning these disparate pathways to the left and right hemispheres, it has been theorised that intuition is rooted in mechanisms that enable the brain to take in subtle patterns and clues from the external environment. The absorption of this information can result in a body of knowledge we are unconscious of acquiring and which cannot be expressed in words, but can be accessed and acted upon 'unconsciously'. This stops some way short of an explanation for intuition, and, so far, there is no absolute experimental proof for a store of hidden knowledge. Yet, as we have seen earlier, there is plenty of evidence that the right hemisphere has an ability to cope with very much more detail than the left. And this detail, in some circumstances, can be brought into conscious awareness.

We also know, from work on patients with amnesia, that memories that are lost to conscious recall (but not to the overall system) can still indirectly influence that person's behaviour. Such studies suggest that our brains possess two memory systems – one for regular memories that can be accessed by our verbal system and one for implicit memories that are off-limits to conscious awareness.

From such work, Guy Claxton, a psychologist from Bristol University, has proposed that we possess a 'smart unconscious' that he calls the 'undermind'. He argues that intuition is evolution's default strategy for solving problems and society, at least Western society, has lost sight of it. To rectify this oversight, he recommends that school children should be encouraged to meditate on problems to enhance learning. From his observations, he has seen that when one allows the brain to soften and draw in other influences that are not immediately thought to be relevant, the brain will find connections. And learning *is* making connections.

This mirrors Schooler's advice for problem solving. He has said: 'Don't try too hard, and avoid words'. His research indicates that verbalisation impairs judgement based on intuition because the brain handles verbal and non-verbal information in different ways and doesn't like mixing the two. To get the most from our intuitive side, he suggests that we don't put pieces of knowledge that are essentially non-verbal into words. For example, instead of describing a face he would advise us to take in the image directly. He has also found that people who are good at intuitive thinking are also good at recognising visual patterns.

This is yet more evidence of two, separate brain pathways. One is based on a linear approach and the other on one that incorporates much more subtle detail. We may ask what is the more efficient system – the intuitive instantaneous way or the slow rational intellectual way?

Transcendent experience

There is much, particularly on the fringes of the 'paranormal', that is unexplained by the current models of consciousness. We suggest that religious experience, intuitive communication and all the other semi-mystical functions can also be related to residual higher function in the right hemisphere. The many approaches for accessing these experiences include religious ritual, meditation, shamanistic procedures, and drug-based rites. Trying to get out of speech mode, leaving the endless chattering voice of the left hemisphere and enhancing the system with various chemicals are all techniques for overriding left dominance and activating the non-dominant side.

Techniques used within mystical and religious traditions are in effect designed to suppress the left hemisphere and reengage the right. Both of these approaches are needed to allow brief windows into the hidden way of being. Unfortunately, for most of us, left hemisphere control is hard to circumvent and it snaps back in all too readily, but the pathways can become easier to follow by continual engagement. However, it is commonly found that the left tries its hardest to erect a barrier of fear that blocks 'progress' into altered states of consciousness. This is to be expected, for the security and indeed identity of the left hemisphere self is being threatened by these approaches.

Techniques for actively engaging the right hemisphere include visualisation, mantra, dream techniques, use of consciousness-changing chemicals, refining nutritional intake and opening to the subtle 'energy' of spiritual locations, be they cathedrals, stone circles or waterfalls. All these will be more effective if right hemisphere suppression is lifted first. We have already seen how restricting sleep can reduce left hemisphere interference, but more usual approaches involve getting out of speech mode and quieting the chattering voice.

By using mind tools, such as focusing on the breath or repeating a simple vocal formula, it is possible, occasionally, to access states in which time implodes and there is no separation or resistance. Fear can drop away leaving profound peace and bliss. These moments of insight or enlightenment feel profoundly real and all-encompassing, but the fact that we have such limited access, and to achieve even this requires dedicated chanting, meditation or prayer, suggests that there has been some sort of system failure. When they do occur, these blessed experiences are often brief and may not be repeated within an individual's lifetime. Is this satisfactory and all we can hope for? If these states are a partial relic of something much more sustainable, perhaps a combined scientific and religious approach could help us regain our 'pre-fall from grace' perceptual heritage. What seems certain is that by removing left hemisphere control one is removing the biggest single impediment to accessing higher states. The door to paradise slammed shut with the establishment of cerebral dominance.

There does seem to be a suppressed awareness somewhere inside us that something is not right – some feeling of disconnection that comes bubbling up from time to time. Could this be because we all are hybrid selves? Though the left dominates, the right is still present – it is not entirely overshadowed. Maybe this is where these feelings come from. One of the basic Buddhist principles is that of 'dukkha' – the unsatisfactoriness of all conditioned phenomena. This two and a half thousand-year-old teaching points to this hybrid nature of our consciousness. There is a fundamental uneasiness in us. The whole basis and motivating factor for religion is fuelled by our striving to come to terms with this internal glitch and to return to an Eden-like consciousness. The religious experience is a coming home. It is not going somewhere new. It is a realisation that this is where we should be all the time. It is closer to us than anything else. It is not out there but from the position of our left brain 'out there' is precisely where it is looking. This is a bizarre conundrum.

A Different Perspective

The human traits we automatically assume are an aspect of high function now reflect our present value system. Civilisation, as we recognise it, is almost evidence of dysfunction. The more we strive for technological solutions, the more of a problem we seem to make. The highest states of mind are not reached by external cleverness. Spiritual tradition recognises that these are to do with something beyond this. It can be argued that the greatest human value lies in these states, not in the cornucopia of distraction that modern society offers. Living simply, in harmony with 'what is', connected to the flow of life and brimming over with intuitive knowledge is to be truly human. In this state, the things for which we usually crave will seem like children's toys. We are dominated by a very distorted view of things.

In the highest state, there is a profoundly intense feeling of just existence with no answers or even questions – an all-encompassing joyful beingness.

The experience of astronaut Edgar Mitchell serves as a good example of something we can all potentially access: whilst staring out into space, he was taken over by a majestic feeling of connectedness. He became aware of something like an enormous force field that connected all people, all their intentions and thoughts, and every animate and inanimate form of matter. Time was just an artificial construct. Everything he had been taught about the universe, and the separateness of people and things, felt wrong. This was not something he comprehended with his usual mind but was an overwhelming visceral feeling. He felt like he had physically extended himself to the furthest reaches of the cosmos. Such states are intense, full of perpetual wonder, but they do not necessarily render us outwardly incapable. Edgar Mitchell continued to carry out his tasks. It is possible to act – to fuel the system that facilitates that experience.

Our rational side has some difficulty with all this. It sounds a bit flaky. After all dissatisfaction fuels progress does it not? What would happen if our lives were not filled with relentless craving? Where would the desire for a Mercedes Benz and the latest electronic gizmo go? It would not begin to impinge. A society based on this deep intuitive knowing would, of course, be radically different from the fearful, desire-based, environmentally destructive society we have today.

SUMMARY ~ψ~

Most scientists today believe our two brain hemispheres have their own specialist functions. An increasing body of evidence however conflicts with this view. It seems clear to us that the right hemisphere has basically everything the left has and had, and that the left hemisphere is a degenerate equivalent of a right hemisphere. Suppression of the right keeps its more outstanding abilities in check, and poor nutritional building materials plus a lack of the optimum 'fuel' also limits its function.

Religious rituals, shamanic rites, dance and music can give us some access to right hemisphere function (whole-being states). This is why these activities have always been important aspects of human culture. But such states may just be the fringes of what an unhindered and fully fuelled up right hemisphere is capable of achieving. Experimenting with sustainable methods of access may cast light on significant latent function and a vastly different sense of self.

However, what we attempt to describe as 'greater potential function' or 'whole-being states' are in fact analyses made by our rational computing system. Stepping out of its mode of operation and viewing experience from the other side could radically alter the way we see, feel and exist. Our sense of self would shift from being dominated by symbols and concepts to an 'experiential beingness' – a state that is inherently difficult to define in linear terminology. This book is, of course, written by and for the left hemisphere!

Conceptually this is such an interesting enigma. Our divided and entangled consciousness system may have resolved this problematic scenario had it not been so obtuse. But as the equipment that generates the sense of self has gone haywire, the solution has disappeared behind a veil of mirrors. To try to sort out this confusion with a piece of equipment that is failing remains a complicated puzzle. The gremlins are in the machine that is producing the sense of self, and the sense of self is trying to figure out the whereabouts of the gremlins when they are right in the machine that is producing them! The first step in any solution, of course, is recognising that there is a problem.

All this has massive implications. The culture we have produced is dominated by restrictive linear systems and, because of this one-dimensionality, we cannot see the overall problem. We have ended up with supermarkets full of food items, which we know are less than healthy but we continue to eat, a population wracked by degenerative diseases and a health service cracking under the pressure. Our children are overweight and under-perform. Relationships and marriages appear less and less sustainable. And there are huge problems with interpersonal and inter-societal suspicion, distrust and violence. Just occasionally too, elements in society come up with flashes of brilliance. There is a lack of wisdom and a general blindness tempered by genuine creative insight – a clear reflection of a dominant dysfunctional system tempered by a suppressed functional one.

CHAPTER FIVE

~ψ~

Fertility and Function

If the whole neuroendocrine model that we outlined in Chapter Three has any validity, we would expect clues that offer support to emerge elsewhere on our palette of human systems. We have already implied that a human system released from the left hemisphere's control could exhibit quite fundamental physiological differences. Digestion, immune function and even structural changes could result from a change in hemispheric controller via, of course, the chemical messengers that are modulated by that controller. In this chapter we investigate the effect that powerful steroid inhibition would have had on our biology, particularly within those systems in which hormones play a key role.

HUMAN FERTILITY ~ψ~

It becomes immediately apparent when we compare human reproduction to that of any other animal that we are stuck with a very inefficient system. 'Just as well' you might say, 'our planet is crowded enough' but this is to miss the biological point.

One essential property of evolution is that it results in exceptionally efficient systems. It is a harsh process. Traits that do not work well are selected out of populations. Living systems that do not fit their environment fail to thrive and, without change, become extinct. Most animals have very efficient reproductive systems that ensure conception with a minimum number of matings. In animals that can breed at any time of the year, ovulation is linked in a direct way to the act of copulation, but this does not happen in humans. As a result, the success rate of matings to conceptions is a lowly 5% while in other mammals it is about 95%. This doesn't necessarily prove anything, but it does beg the question of how, in evolutionary terms, such an inefficient system could have come about.

The whole process of ovulating, menstruating and rebuilding the uterus every month also appears extraordinarily wasteful – particularly as it occurs even when there has been no sexual activity at all. Although there are some parallels with other animals in which there are periods of oestrous, the human system is unique in its wastefulness of resources.

Another anomaly of the human reproductive system is menopause. The loss of fertility long before death is virtually unique to humans. We can see from hormone replacement therapy that simply putting back a few powerful hormones, even in their synthetic form, can restore the reproductive system to something approaching a pre-menopausal level of function. Could this be a clue to what has happened in humans?

All these anomalous factors have left us wondering how the female cycle, as we know it today, could have evolved? What is the evolutionary advantage of monthly ovulation and heavy external bleeding? Something in all this is highly unsatisfactory.

Orgasm

There are questions around the function of the human orgasm too. What is it for and why did it evolve? In the male, orgasm is coupled with the release of sperm. Why is this not paralleled in the female? Perhaps in our pre-fall state the female orgasm was a releasing mechanism for the ovum.

We know of course that there is a correlation between male orgasm, the function of the testes and the release of sperm. Foetal development of males (see Chapter Three) shows us that males are not much more than hormonal variants of females. Genetically males and females are virtually identical. Thus it would seem likely that orgasm should have a parallel function in both sexes.

If the female orgasm evolved as a mechanism for releasing the egg, what would this imply and why doesn't it work now? In males, there is a psychological and neurological response (the feeling) connected with a mechanical response (muscle spasm). Perhaps in the female something has occurred to break the link between the orgasmic responses that we do see and the release of the egg.

In the previous chapters, we have proposed a variation on the standard model of human evolution that postulates a significant change in our neuroendocrine system. Such a change would have affected many aspects of our psychology and physiology. The reproductive system is perhaps the most hormonally sensitive of all our physiological functions. The smallest shift in hormone activity can induce changes in structure and function. Thus, if we return to our hypothetical fruit-eating population of proto-*Homo sapiens*, we need to ask what would have been the effect on the reproductive system of a diet rich in steroid-suppressing chemicals? What would have happened to fertility when the proposed powerful internal pineal loop kicked in? Then, having adapted to such an altered hormonal regime, what would have been the result when a rapid loss of steroid inhibition exposed the product of a unique evolutionary variation to a far higher level of steroid activity. It is impossible to significantly alter hormone activity within the body without having an effect. Some change in our reproductive system would have been inevitable.

To clarify these issues, we can divide our development into three evolutionary periods: the ancestral period, the period of steroid inhibition with rapid brain expansion, and the subsequent period in which this steroid inhibition was progressively lost.

THE ANCESTRAL PERIOD ~ψ~

It is reasonable to assume that early hominids and primates had similar reproductive systems with repeating oestrous cycles that provided a window of sexual receptivity and ovulation. Mating (at least successful mating) would have pretty much been confined to this receptive time.

The hominids in the forest would have eaten a largely fruit/leaf/flower/shoot diet. Some groups or species may have specialised more on one aspect of the diet than another but in all groups, fruit would have been a much-favoured element. As we have seen, this diet contains chemicals that mimic animal oestrogen, and it is acknowledged that these phytosterols/hormone mimics can have a significant effect. They do this by adding to the pool of oestrogen but, as they are less potent than their human counterparts, they dilute and weaken overall oestrogen activity. They can further reduce oestrogen activity by blocking oestrogen receptor sites and even mopping up oestrogen molecules themselves. Phytosterols therefore will suppress oestrogen levels in the body.

This can markedly affect fertility. Recent investigations into the use of soya-based concentrates, such as baby foods, have revealed that the amount of oestrogen-like substances that they contain can equate in strength to a couple of birth control pills a day. It is thought that the sustained high intake of soy isoflavones (a type of phytoestrogen) amongst Japanese women may account for their menstrual cycles being on average 32 days in length – about three days longer than those of western women. Another study found that sheep became infertile when they were allowed to graze on a monoculture of clover – it too was rich in an isoflavonoid phytoestrogen (formononetin).

If our ancestors had a daily intake of two to three kilos of fruit and shoots rich in these compounds, it would have had a significant effect on the reproductive system (and indeed on the functioning of the mind and body as a whole). The bioflavonoids and related phytosterols found in a primarily fruit/vegetarian diet would have been powerful enough to modify our ancestor's internal hormonal balance.

THE PERIOD OF STEROID INHIBITION ~ψ~

We have seen that the period of rapid evolution that mankind went through was fuelled by external and subsequently internally produced chemicals which markedly suppressed steroid activity within the body. This would have caused many changes in structure and function but potentially none would be greater than the change to a reproductive system that we know is so hormonally sensitive. Thus a reproductive system which had slowly evolved over millions of years was, in evolutionary terms, 'suddenly' subject to increasingly powerful steroid inhibition. What could have emerged to accommodate this new regime for it is certain that, if this scenario is correct, it would have to change?

It is possible that with powerful external and internal suppression of steroids, fertility could have over time become very low but immunity and health very high. Our reproductive system would have adapted to the high level of steroid inhibition to compensate for low fertility resulting in a unique mechanism able to cope with this unique scenario.

Flavonoids and the female cycle

It is evident that our present day reproductive cycle can be altered by diet alone. In baboons (a species that does have menstruation) it has been found that menstruation stops when females are fed a vegetable only diet. We know of course that the human female cycle can be changed by overt methods like contraceptive pills but, as we have already seen, even plant hormones, such as Soya bioflavonoids, can alter the monthly cycle. Leslie and Susannah Kenton in 'Raw Energy' have commented upon this effect:

'Women on an all-raw or high raw diet often report that menstrual problems such as bloating and pre-menstrual tension and fatigue greatly improve after two or three months. For some of them the improvement is so dramatic that they are not aware of their periods until they arrive. Heavy periods become lighter – a period that ordinarily lasts six or seven days can be reduced to as few as one or two. In some women, particularly those who did not eat meat, dairy products or large quantities of nuts, periods even cease altogether.'

The absence of periods in women with diets rich in fruit and vegetables has been linked to increased levels of carotene. A team of gynaecologists at Rutgers University in New Jersey investigated a group of women who exhibited both carotenemia (a change in skin tone due to large amounts of carotene absorbed from carrots and other vegetables) and amenorrhoea (a cessation of periods). They wanted to ascertain whether there was a link between the two conditions. All the women were in excellent health and were not adversely affected by these complaints. The team found that when carotene was excluded from the diet their menstrual period returned. When they reverted back to their original diet they became amenorrhoiec again.

It is not clear from this study whether fertility was affected. It is possible that ovulation continued without the need for menstruation. There are cases in which women have become pregnant even though they have ceased to have periods. Taking the pill all the time (not breaking it for a week each month) also stops periods but occasionally pregnancies can still occur. Menstruation is not fertility, though they are of course linked in our present state. Eating a diet rich in fruit and vegetables can stop menstruation and ovulation. However there may be a finely balanced point in which ovulation may still occur without the need for menstruation. Whether this is just an occasional aberration or something more is in need of clarification.

British gynaecologist C.A.B. Clemetson has also investigated changes to the female cycle caused by chemicals within fruit and vegetables. He became interested in the subject when a young Italian woman told him that she reduced her excessive menstrual bleeding by sucking lemons – apparently the local folk cure for the problem. Clemetson found that citrus bioflavonoids in the blood could indeed reduce menorrhagia (heavy periods). Doses of the natural chemicals taken with vitamin C for three or four months significantly reduced excessive bleeding in most of the women who participated in the trial. And many of these maintained their lighter periods when they acted on his recommendation to eat three oranges a day. If you are tempted to try this, remember to eat plenty of the white pith, for it is the pith that is richest in bioflavonoids.

Several of the bioflavonoids, found in fruit and vegetables, mimic the female sex hormone, oestrogen. Like oestrogen, the bioflavonoids strengthen the fragile capillaries that line the walls of the uterus. It seems that at the point of the female cycle when oestrogen levels are low, i.e. during menstruation, the bioflavonoids take over the task of strengthening the capillary walls that would otherwise break down. This helps to reduce menstrual flow. It is the cyclical fall in oestrogen that brings on the breakdown of the walls of the uterus and the subsequent bleeding. If oestrogen levels do not fluctuate (but remain either high or low) menstruation does not occur.

We can see then how bioflavonoids can ameliorate the effects of the cyclical rise and fall of oestrogen levels and that a diet high in carotene can actually stop the process altogether. It is not a great leap therefore to suggest that at some time in our evolutionary past, the steroid inhibiting compounds taken in from a fruit-rich diet could have stabilised oestrogen at a low activity level, resulting in a cessation of monthly periods. By building this missing link into the model of human biochemistry, a different picture emerges that seems to provide some explanation for the many anomalies that are apparent in the human reproductive system.

Indeed the whole fertility cycle as we know it today may be the result of a period of powerful steroid inhibition followed by the breakdown of this effect. Menstruation may thus be a symptom of hormone imbalance. However, the fertility cycle is extremely complicated. Rafts of different hormones, including estradiol, nor-adrenalin, luteinising hormone and gonadotropins, together with associated neurochemicals interact to regulate each other's activity. Altering any one element would have many knock-on effects; and the scenario we have proposed would have altered many of these elements. Because of the complexity of the system it is difficult to track the full ramifications of the different interactions and feedback loops. We can speculate however on the most likely outcomes of this model.

Ovulation and orgasm

Basically we think that the combination of the dietary hormones and increasing internally produced melatonin acted like a contraceptive on the archaic oestrus cycle. The hormonal cues and triggers that regulated the oestrus cycle were increasingly dampened to the point where ovulation became less frequent and the cycle itself began to stall. This scenario would have produced strong selection pressures to find an effective solution. If reproduction were becoming more difficult, any changes that would have enhanced fertility would have had more chance of being passed to the next generation. A chemical/hormonal trigger produced during copulation, sufficient to tip the balance from effective contraception to ovulation, would have been such a change that could have enhanced fertility.

If the flood of steroid inhibitors held the female reproductive cycle in a state of suspension, something would have been needed to induce ovulation. And, for maximum efficiency, ovulation would need to be induced at the time of mating. A physical mechanism, linked to a psychological/neurochemical one, may then have been the key to releasing the ovum in response to sexual stimulation. Orgasm, therefore, could have been a central part of this mechanism, acting for the female in the way that it still works for the male.

Intriguingly, it has recently been found that human semen contains hormones, such as follicle stimulating hormone (FSH), luteinising hormone (LH) and estradial, which are known to induce ovulation. FSH actually causes the egg to ripen and burst out of the ovary. In comparison, these hormones are either not present (LH) or occur at much lower levels (FSH) in chimp semen. Roger Gosden, at the Center for Reproductive Medicine and Infertility in New York, says he and others are 'mystified' by human semen's composition, but it makes great sense if mating stimulated ovulation.

As the model we have proposed suggests ever-increasing levels of melatonin, any solution to this fertility problem would need to keep pace. An increasingly powerful neuroendocrine response, linked with copulation to trigger the now stalled oestrus cycle, may have been the result. Specifically this would mean sufficiently powerful and sustained orgasmic states would have been needed to induce ovulation.

The human fertility cycle is usually portrayed as being regulated by a small number of key hormones and some neural feedback mechanisms: Cyclical increases in levels of steroids induce a surge of luteinising hormone that in turn triggers ovulation. Recent research has found that the picture is more complicated, with the brain itself playing a bigger role, particularly in regard to the release of luteinising hormone.

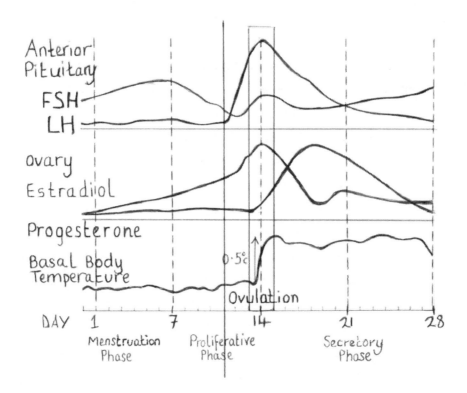

Fig 6a: This diagram illustrates the typical hormone patterns in the human menstrual cycle. The line at day 11 represents the hormonal balance that would have been in stasis in our proposed 'stalled cycle' until orgasm induced ovulation.

This neural regulation of luteinising hormone (LH) may be important. The part of the brain that helps to regulate fertility hormones, including LH, is the hypothalamus. The hypothalamus is also involved in excitement and has direct connections to areas of the brain concerned with pleasure. Thus pathways seem to exist that link orgasm (excitement and pleasure) to neuroendocrine changes that induce ovulation. In the past, then, it is certainly tenable that orgasm may have stimulated ovulation but for this to occur, orgasm may have had to be intense and sustained.

In most circumstances today, the hormonal effects induced by orgasm are not powerful enough to induce ovulation, but some hormonal changes do occur. Orgasm is known to elevate LH and oxytocin. Oxytocin in turn increases steroid levels that can further stimulate LH production.

There are physical affects of orgasm that are consistent with this model too. When a woman achieves orgasm just after the man, the cervix dips, scooping into the pool of sperm deposited near it. Female orgasmic contractions can also enhance the passage of spermatozoa towards the fallopian tube, thus increasing the likelihood of conception. Significantly, this is under hormonal control. Oxytocin is the hormone that stimulates smooth muscle tissue in the wall of the uterus and is associated with promoting labour and delivery (oxy = quick and tokus = childbirth). But circulating concentrations of oxytocin also rise in both males and females during sexual arousal and peak at orgasm. There is evidence that it stimulates, in males, smooth muscle contractions in the walls of the sperm duct and prostate gland, and in females, contractions in the uterus and vagina that promote sperm transport towards the uterine tubes. There is also evidence that the fluids released during female orgasm can also help sperm reach their goal. Female orgasm is then, at some level, still linked to mechanical function, but should it go one stage further and release the ovum too?

It would have been an absolute necessity for a powerful and efficient system of reproduction to be established to overcome the effects of increasing steroid inhibition. The increased inhibition may have directly impinged upon the ovulation mechanism. In addition, if there were a reducing window of sexual receptivity, the efficiency of the system would need to be maximised to ensure the continuance of the species. The primary role of orgasm in the female could therefore have been the induction of ovulation – a very efficient process. Orgasm may have had a very important secondary role too, for in a being that had a very pleasurable natural state of consciousness, a reward may have been necessary to encourage sufficient sexual activity for procreation. And orgasm may have become such an intense reward because the basic state of the mind of man 'before the fall' could have been of a different order than it is today. We may have been blissed-out on wild figs, bananas and beingness.

If there were a link between orgasm and ovulation, this would of course invariably lead to conception. So is there any evidence that links orgasm to conception today? We would not expect to find an obvious correlation, but there is some anecdotal information available. There are cases in which it appears that particularly intense and/or prolonged orgasm have resulted in conception at 'safe' periods in the cycle. It has also been observed that prolonged intercourse and/or orgasmic states can induce slight bleeding. This may not occur immediately but it does within a day or so. It would seem therefore that the bleeding is not a result of tissue damage but would be similar to the slight bleeding that can occur with ovulation. 'Susan X', a client of hypnotherapist and author David Pedersen, relates how she had the most colossal orgasm she had ever experienced with her extramarital lover. This was followed by almost continuous intercourse for the next three hours, throughout which she maintained a huge sexual plateau that produced multiple orgasmic peaks. The following day she started what she thought was her period but it later emerged that she actually became pregnant as a result of this conjugation.

It is widely known, though rarely openly discussed, that for many women their ability to orgasm at all, let alone maintain a highly intense state for extended periods, seems to be a difficult area. In contrast, the rapidity of the male orgasm has been an issue for as long as anyone can remember. However there are exceptions: some female orgasms are much more intense than others. These tend to come about after a sustained level of sexual activity. And sometimes they can be accompanied by a female ejaculate. It has also been noted that high levels of testosterone and oestrogen are found in the bloodstream during these particularly intense orgasms. As we have already noted, elevated levels of these hormones stimulate LH production. This strengthens the possible link between ovulation and intense orgasm. More research on this unusual function is necessary to answer all the questions posed by the mere existence of these states, but we are left wondering how could these high orgasmic states, and their accompanying physiological alterations, have evolved if they were not at one time the norm? They couldn't have been hanging around latent, just waiting for sexual researchers to discover them.

This part of our fertility theory intersects with our consciousness model, for 'high orgasm', as we may term it, could be primarily a right hemisphere function. The very intense states may no longer be available to the left hemisphere, and its dominance may be one reason why orgasm is suppressed in some women and difficult or even impossible to achieve. A change in dominance between the right and left hemispheres may also explain why sexual activity, and particularly orgasm, can take us away from our limited ego state into blissful transpersonal states of being.

We have already noted recent research that suggests that the brain may play a direct part in regulating fertility hormones (neural regulation of LH). The 'pre-fall' brain may have had an even larger role. Orgasmic response and subsequent hormonal release may have been many times more significant. It is likely that left hemisphere dominance is inhibiting these effects today.

Males, diet and libido

So, according to this theory, females have suffered structural and functional changes to their reproductive system, but what about males? With steroid inhibition, testosterone levels or activity would be at a lower general level than they are today. As testosterone is a primary factor in libido (for both sexes), we can speculate that at this critical time in our evolution, male libido would also be of a different order and it would have been more difficult to reach orgasm. (Almost the reverse of the situation today.) Thus a sustained amount of sexual activity would have been needed in both males and females to achieve both orgasm and a release of sperm and ova. If a male climaxed without a corresponding female orgasm, the ovum would not be released and conception would not occur. Male and female sexuality thus had to match for successful reproduction, and the reward was an intense feeling of bliss that was greater than the background benign state of consciousness. This scenario provides a realistic explanation for the presence of orgasm in humans. Something unique must have occurred somewhere along our line of descent for the strange anomaly of pleasurable orgasm to become such a strong feature of the human bio-system.

Diet certainly affects libido. In one touching account, alternative writer Robert Wilson reminisces about a natural-food crusader he once knew. This fellow regarded most food fanatics as hopelessly corrupt compromisers; only *he* had the correct 'natural' diet, which consisted entirely of fruit, nuts and *uncooked* vegetables. When asked; "what was the greatest single benefit he got out of that regime," he replied at once that it solved all his sex problems. He expanded on this beguiling statement: "I hardly have any sex drive anymore," he said, unabashed, even a bit proud. "I don't *need* women the way I used to. I'm free. No problems in *that* area at *all*." Many followers of raw food and fruitarian diets have noticed a related change from an obsessive psychological desire for orgasm to a more expansive sensuality.

We are addicted to sex but could part of this be an addiction to the feeling we get from being momentarily free from our ego-based, fear-ridden, left hemisphere sense of self. Is the sexual drive, which has reached obsessive levels in our society today, a result of a striving to regain something that in our deepest being we know we have lost? At some fundamental level we know there is something more to the sexual experience but, because we don't know where to look, the desire becomes attached to the whole raft of sexual expression from glossy car adverts to the darkest depravity. This distortion arises from a human mind system that has become disconnected from the true needs of the body and from a more balanced, intuitive and complete side of us. The sexual obsession, which is everywhere in western society (sex sells everything), compensates for the very inefficient reproductive mechanism we are left with today. In fact it has over compensated by a very large factor. We are well out of balance – 6000,000,000 sex-obsessed humans to date and still copulating!

THE FALL ~ψ~

The last of our direct ancestors moved or were forced out of the forest some 200,000 years ago. This eviction from 'the Garden of Eden' necessitated a change of diet, which initiated an increase in steroid activity. This would have caused a corresponding decrease in the production of melatonin and pinoline. These two factors would have profoundly affected many areas of the body's biochemistry but particularly the reproductive mechanism.

Melatonin has been used, in trials at least, as the basic ingredient of a contraceptive pill for it has been found to stop (what is regarded as) the normal female cycle. This stalled cycle would have been the norm before our ancestors left the forest. Higher melatonin levels resulting from a highly boosted pineal pump would have produced a system that remained in stasis between copulation-induced ovulation/pregnancy events.

According to this part of our hypothesis, when our 'pre-fall' ancestors eventually reached puberty (at a later age than today because more melatonin in the system would have inhibited the activity of the sex hormones), the female cycle that we are familiar with today would not have occurred. Within the maturing female, the uterine system would partially build in response to her maturing hormonal regime, and from then on it would be set to go. With a diet high in natural oestrogen-like mimics and the pineal pumping high levels of melatonin, her uterus would have been held in partial readiness for conception with no regular ovulation and no monthly uterine breakdown. The system would therefore reach a stasis, similar to the regime under the control of contraceptive hormones, with a partially mature ovum and a partially built uterus. Ovulation would have been held in check by the very different hormone regime but it would have been held close to release all the time (Similar to day 11 of a typical cycle now – see figures 6a and 6b).

It would have only taken powerful sexual stimulation for the whole thing to kick in. Sustained and intense orgasmic states would have been enough to temporarily alter the delicate hormonal balance (helped by the hormones in semen) by inducing higher levels of

LH, steroids and others hormones that would in turn allow the egg to be released at the time of mating. This represents a highly efficient fertility/reproduction system without any needless waste.

With the loss of the steroid inhibition the brakes came off. The system started to behave strangely. It was no longer held waiting in readiness. Before the fall, as puberty set in, internal hormone production merely had to raise the background oestrogen levels to a point where differentiation, leading to sexual maturity, took place. Then, as levels receded, the sustained external oestrogen mimics kept the uterus stable and intact. But now the female hormones do their work without the background level of steroid inhibition to act as a buffer. Once the 'build a sexually mature human' phase is completed, the newly mature system is exposed to a much higher background steroid activity.

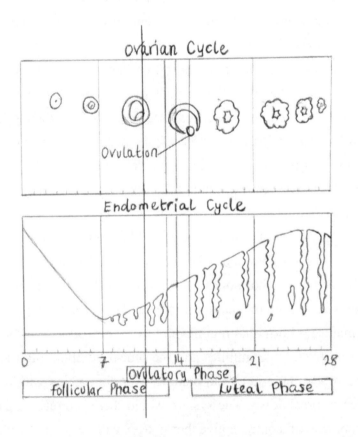

Fig 6b: This diagram also illustrates the typical menstrual cycle. In the normal cycle it is apparent that the uterus lining is only partially thickened before the hormonal changes that occur with ovulation. Even this degree of thickening may be due to excess steroid activity. Again a line has been added to give some idea of how an oestrus cycle in stasis may have looked: a rise in key hormones brought about by intense and sustained orgasm being enough to trigger ovulation that in turn bring further hormonal changes.

So now rising steroid levels trigger a surge of LH that in turn induces ovulation without sexual activity. Ovulation without fertilisation triggers a number of feedback mechanisms similar to those that used to regulate the archaic oestrus. A new cycle is created that still allows incidental fertilisation but the process is inefficient and damaging.

As it is today, high levels of steroids promote the uterus lining to grow rapidly (possibly to excess) and, within a few days, this same hormonal imbalance also induces ovulation. More hormones released at ovulation further stimulate growth of the lining. However without fertilisation further hormonal changes, including a reduction in steroids, eventually lead to its breakdown. Steroid levels then begin to rise again and the cycle repeats. In contrast, in our 'pre-fall' days, ovulation itself would have induced the hormonal changes that would have brought the lining of the uterus to a state in which it was ready for implantation. And this would only occur after a successful mating.

Prior to the loss of inhibition, the rebuilding of a receptive uterus would have occurred once or perhaps a few times at puberty and thereafter only following a pregnancy and on the rare occasions when a copulation/ovulation failed to result in conception. There would have been long windows between rebuilds for, after a pregnancy, it is likely that the inhibitory effect (on the fertility system) of breast-feeding would have been stronger under a modified steroid regime. (Even now the hormonal changes brought about by regular breast-feeding will inhibit conception 99% of the time in the first year and about 97% in the second.)

Chimps on average have seven years between conceptions and so a similar period in humans may be 'more natural' than the relatively short periods we are familiar with today. Thus with relatively few conceptions during a female's life and consequently few rebuilds, the system would be well able to cope. It cannot however cope with a rebuild every month coupled with the hormonal 'roller-coaster'. The consequence is menopause – an anomaly virtually unique to humans. The system collapses in on itself. It cannot do any more. The high steroid regime wins in the end and women become a bit more like men.

A major investigation into the connections between female fertility, bioflavonoids, a raw food diet and melatonin is not only long overdue but would also test this thesis. If it was found that a low steroid regime could restore some stability to a system which appears to have become one of runaway self-destruction, at least half of the world population could benefit enormously.

Whilst we are looking for symptoms that might suggest we are at the wrong end of the re-emergence of a high steroid /low melatonin regime, it may be fun to turn our attention to the perennial problem of hair. For this discussion, and indeed this whole chapter, we need to keep in mind that the steroid regime, which is regarded as more or less normal, is in fact, within the context of this model, abnormally high.

Male pattern baldness is accepted as a natural consequence of ageing. Baldness however is not deemed acceptable to most middle aged men inspecting their receding hairlines in the mirror every morning. A full head of shiny hair is a symbol of youth and attractiveness and we want to hang on to it for as long as possible. Hair is power. Throughout history, long hair on warriors has been a symbol of strength, and even the words Caesar, Kaiser and Tsar mean hairy in their original form.

As a consequence of these feelings about hair, a whole host of treatments have been devised to counter its loss. One of the earliest documented cures goes back to around 400 BC. The Greek physician Hippocrates recommended rubbing the scalp with an ointment made of opium mixed with wine, acacia juice and the oil of olives and roses. For severe cases, a paste of cumin, pigeon droppings, beetroot, nettles and horseradish was prescribed instead. Many modern approaches are similarly bizarre and most have only minimal effect, but it is significant that there are now hormonally-based remedies that do partially reverse the process.

The hormone most often fingered for crimes against hair is the androgen dihydrotestosterone (DHT). This can cause hair follicles, of both men and women, to become dormant. Excess DHT can be combated with anti-androgens. The very fact that we can respond to such treatment again shows how much plasticity there is within the human system: that balding can be treated with something that reduces the activity of the testosterone group of hormones is a further indication that we are suffering from excessive effects of these hormones. As an aside, it is also worth noting, in the light of the previous section, that one of the effects of taking hormones for the 'hair today, gone tomorrow' problem can be a loss of (standard ideas of) libido.

One could also ask what the evolutionary advantages of early hair loss are – particularly for those savannah dwelling and evolving humans? How could this trait of balding have possibly come about? That it is an anomaly, thrown up by an archaic change in steroids, makes more sense than any existing theories. Perhaps those who still think that man evolved on the savannah would like to explain the advantages that hair loss would bring to a male on the sun-drenched plains of Africa.

With increasing age, many women also experience head hair thinning along with hair thickening on the upper lip and sometimes other areas of the face. As we value a glamorous idea of beauty more highly than just about anything else, this is seen as a problem. Female pattern baldness like male baldness involves genetics, vitamin/mineral uptake and stress but the key component of the problem is again hormonal balance. As well as excess of (DHT), balanced thyroid hormone production is also critical; hypothyroidism results in coarse lifeless hair which easily falls out, while hyperthyroidism causes soft, thinning hair. In women at menopause, the ratio of androgens to oestrogen changes and this is often the trigger that leads to increases in facial hair and head hair loss.

These problems, in both men and women, are almost certainly compounded by the reduction of melatonin production that occurs as we age. We have seen that steroids, or at least their activity, can be inhibited by bioflavonoids and melatonin, thus it is possible that diet as well as activities that boost melatonin production could ameliorate these negative effects. All this suggests that we have a body system today that is abnormal. With a much lower level of steroid hormones none of these strange things would happen.

Stress can also cause hair problems. Cortico steroids (the stress hormones that are closely related to testosterone) are enough on their own to activate facial hair follicles in women. It has been noted that, in stressed middle-aged women with a facial hair problem, dealing with the stress alone can have an effect. For example, changing the consciousness system from that of stress to one blissed-out on love can actually stop the unwanted hair growth.

Some psychologists think that dominance in females (and probably men as well) is a trait closely associated with high levels of testosterone. And some research shows that female students, managers and professionals secrete above average levels. They may not all be hairy but hirsutism is an extreme 'hairy condition' in which, in response to abnormally high levels of androgens, there is excessive growth of hair on the face or body. Usually it is only regarded as a problem for women – hairy males are more or less socially acceptable. And in women it can be accompanied by other changes such as balding, deepening of the voice, increased muscle mass, loss of breast tissue and acne. These masculine traits are typical of our (cruel) characterisation of 'dominant' woman. New research from the University of Auckland has found that even the way women view themselves is directly related to the degree of testosterone they secrete: women who actively thought that they had dominant traits actually secreted higher levels of testosterone. So like the example of the woman in love, it seems that what we do with our consciousness can affect our biochemistry, and this in turn can feed back into our physical structure as well as our behaviour. What a complicated system. No wonder we have such a problem with relationships, wars and, of course, hair.

The link between acne and abnormally high levels of testosterone has a bearing on another anomaly of our human system that can also be explained by too high levels of these pesky steroids. Sebaceous glands in humans seem to have gone crazy. These glands are an appendage of hair follicles, and they produce an oily substance called sebum that helps to keep the fur of mammals sleek and waterproof. In a naked ape, it would be reasonable to expect these glands to have become largely vestigial, but the opposite seems to have happened. The sebaceous glands of African apes are found scattered all over their bodies, but they are small and few in number. In man, they not only are more numerous but also much bigger, especially on the face and scalp. Sebum does not seem to have any usefulness to us in our hairless state – it is not needed to keep the skin moist or supple. In fact, if anything, it causes problems. Sebum provides a breeding ground for dandruff and dermatitis, and is associated with pimples, black heads and inflamed nodules.

The sebaceous glands do not begin to operate until puberty. This is why adolescents have such a sudden problem with acne and related skin blemishes. Sebaceous gland activity is a response to sex hormones. It is likely that it is an anomalous response. We don't need sebum except perhaps to add gloss to our head hair, and, as children, we don't produce much of the stuff at all. Thus hugely active sebaceous glands have not evolved in man for a purpose. Their ebullient activity is rather another indication of our anomalous levels of steroid hormones. Under a much lower steroid regime, the sebaceous glands would not develop in the same excessive way.

SEROTONIN, TESTOSTERONE, DEPRESSION AND MURDER ~ψ~

Studies of individuals who display psychopathic behaviour and have been convicted of the darkest of crimes show that they suffer from low levels of serotonin and high levels of testosterone. This is in effect, according to our theory, the extreme end of what all of humanity has been exposed to. The combination of low serotonin with high testosterone appears to be linked to violence, the willingness to kill, uninhibited aggression, and a detached unemotional mental state with a total absence of compassion. Such psychopathic individuals have a greater sense of disconnection from others and a greater sense of isolation. We can surmise that they have even less of the biochemistry to fuel the sense of connection to all life, and left hemisphere dominance is more complete.

In instances when chemical or physical castration has been carried out on the worst offenders (which in cases of sex attacks and serial murders are virtually always men) the incidence of re-offending has been reduced by a massive factor of twenty. Reducing testosterone levels thus stops the criminal behaviour almost entirely. This is a staggering result. Just by changing one or two core biochemical pathways, behaviour can be changed. There may be a massive web of chemical interaction that makes up our behaviour patterns but this work shows that testosterone has a disproportionate effect; an imbalance can lead to violent results. (There is a direct connection between the levels of neurotransmitters like serotonin and steroid hormones like testosterone; more steroids lead to less neurotransmitter activity.) If, as we believe, we are all suffering from too much testosterone activity, this connection between testosterone and violence has enormous implications for society.

Recent studies suggest that antisocial conduct may be linked to diet too. Increasing the levels of vitamins, minerals and fatty acids in the diets of young, imprisoned offenders reduced the number of disciplinary offences committed by over a quarter. Significant infringements of the rules, including violence, fell by 37% when supplements were given to a trial group. Though this research has come as a surprise to many people, it is what we would expect if our thesis were correct. Bernard Gesch, who conducted the work whilst he was at Surrey University, has correctly pointed out that nutrients are crucial ingredients in the biological processes that produce brain transmitters, like serotonin and dopamine, which affect mood. As a junk food diet, free from fresh fruit and vegetables, is becoming more prevalent, this evidence suggests we are sowing the seeds for a much more disruptive and violent society.

Furthermore, research on monkeys reared in isolation has found that, in comparison to those benefiting from a normal interactive upbringing, they were less socialised and grew into more aggressive adults. These monkeys had less serotonin activity. Linked to this, there is some evidence that suggests that touch, particularly a mother's touch, is very important to babies and young children. It can affect life long levels of serotonin. Perhaps a society that isolates its children into cots and crèches is storing up more and more problems. We should all ask ourselves – is this normal primate behaviour? A lack of touch in early childhood may make depression in later life more likely. It has been stated that depression will soon be the second most widespread medical condition of all. We also know that the prescription and consumption of Prozac is increasing all the time. Does this not indicate that something has not only gone seriously wrong with our internal levels of these chemicals but also that it is getting worse?

TESTOSTERONE, SOCIETY AND CONSCIOUSNESS ~ψ~

As boys develop there are a number of stages during which the levels of testosterone markedly increase. It all begins in the uterus. At about the eighth week of pregnancy, the Y chromosome turns on testosterone production and initiates the formation of testicles. At fifteen weeks these are functional, in that they make extra testosterone, which steers the progression of masculine development. Testosterone levels drop considerably after birth, and during the toddler stage levels stay fairly low. During this phase the behaviour of girls and boys are similar but a change occurs around the age of four; testosterone levels in boys double and their behaviour alters considerably. Boys become boys, interested in action, adventure and boisterous play. At five, testosterone levels drop again by half, which restores some calm to their lives though there is enough of the steroid in the system to maintain an interest in adventure and exploration.

At puberty, testosterone levels increase to around 800 times the amount that boys experienced when they were toddlers. This brings on all the familiar structural developments but it also profoundly affects their sense of self. As well as a general restlessness and the presence of strong sexual feelings, this is the time boys tend to only communicate with grunts. Testosterone is such a powerful substance that it profoundly affects not only male behaviour but also consciousness at a very deep level. It affects what males do and how they think of themselves.

This is illustrated very clearly by an experiment on the dynamics of monkey society. When a lowest ranked male monkey was given a testosterone boost, he challenged his immediate superior and won. He then tackled the next in line and also won. Within twenty minutes, despite being small, he had successfully challenged the top monkey and had become number-one. His high status didn't last. When the testosterone was used up, he was knocked back to his former position.

An abnormally high level of testosterone sometimes occurs in girls. Congenital adrenal hyperplasia can give a girl excess testosterone in her mother's womb. Levels drop back to normal after birth but there is a residual affect. These girls show enhanced athletic skills, prefer male playmates, 'male' toys and like to dress in a more masculine way too.

Just this one steroid has a huge effect on human life. It is very probable that at some stage in our human evolution, when we depended on fruit as our major source of nourishment, testosterone activity would have been lower. The 'fruit effect' and the melatonin boost would have buffered its influence. This is not to negate the hormone's crucial role in masculine (and feminine) development, but back then it wouldn't have had such an overbearing influence, and it wouldn't have caused the subtle structural damage to the neural system.

Testosterone is a key hormone in neural development and body function, which is why restoring its activity to a more natural level is so important. As it is now we could say that society runs on an excess of testosterone. The male influence is hugely dominant. The monkey craving to get to the top of the tree fuels activity. It is older men who control world society.

While testosterone levels in men over forty do drop, so does the production of melatonin that buffers its effect. In addition, the overall reduction in pineal activity will exacerbate left hemisphere dominance as the chemicals the right hemisphere needs and can utilise more effectively than the left decline. Thus by the time men get to middle age the high level of testosterone has already done its damage; the consciousness system becomes dominated by fear and from this stems the obsessive need to control. Our over-arching need for security means we elect old male politicians (or they take control by violence) who spout doctrines of further control and restriction. At some level we know the system is crazy and wrong but we cannot put our finger on the fundamental fault line. It lies however in a society that is burdened with individuals who throughout their lives suffer from an unnatural level of testosterone. The world is dominated by control ridden, power hungry, older men (probably amongst the most hormonally damaged members of our society) who have set up structures to facilitate their schemes. Although the merits of these patriarchal structures could be called into question themselves, if they were filled instead with 25 year old women, a very different world would emerge. It is almost inconceivable that their decisions would lead to the ever-worsening levels of environmental destruction, the brutal outrage of war and the threat of nuclear holocaust.

Society has genius and dysfunction, much like the mind. Society uses its genius side to come up with brilliantly clever discoveries like nuclear fission and the elucidation of DNA, but the final decisions of how these are used are left to those in control, not the geniuses who devised them. Thus we end up with nuclear bombs and the nightmare possibility of cloned humans. All this is an intriguing reflection on the mind itself. The right hemisphere may be responsible for insight and innovation but it is the fear-based left hemisphere that has the control and makes the final decisions.

MELATONIN, THE PINEAL AND SEX ~ψ~

In medical cases in which damage, such as a tumour, blocks the function of the pineal gland in children, puberty is rapidly initiated. The high level of melatonin found in children is therefore implicated in the suppression of sexual development. When levels drop, puberty begins, and in girls ovulation commences. It is also known that melatonin regulates many hormones including those that regulate the menstrual cycle. Fluctuations in melatonin levels stimulate the pituitary gland to release luteinising hormone, oestrogen, prolactin, oxytocin and follicle stimulating hormone (which regulates the production of sperm in men and stimulates the maturation of ovaries in women). Thus we can see that melatonin and the pineal play crucial roles within the human fertility system. Anything that alters the balance of pineal function will affect fertility.

All this has raised the possibility of using melatonin as an oral contraceptive. Endocrinologist Michael Cohen, formally of Dijkzigt University Hospital in Rotterdam, has discovered that high doses of melatonin, when combined with the female hormone progesterone, can block ovulation. Because of questions concerning the safety of the oestrogen-based contraceptive pill (oestrogen has been shown to increase the risk of certain forms of cancer), Dr Cohen set out to develop a new oestrogen-free pill. His research resulted in a contraceptive pill that contained a dose of 75 mg of melatonin with a small amount of synthetic progesterone. During trials, carried out in Holland, over two thousand women took this pill for over three years. It proved every bit as reliable as the standard pill and what was perhaps just as remarkable was that the women reported no unpleasant side effects – no headaches or bloating, both of which can occur with the oestrogen pill. In fact the women in the trial reported a generally heightened sense of well being.

This is very interesting from the point of view of our hypothesis. It is significant and reassuring to see that even when melatonin was taken in very high doses, it appeared completely safe. A trial that raises melatonin levels to such a high degree, and results in not only no negative side effects but also positive ones, certainly doesn't contradict the idea that at some point in our past our pineal glands may have pumped much more melatonin. Additionally Michael Cohen found that the high levels of melatonin did not, as one might expect, increase drowsiness. Melatonin taken in low doses acts as a sleep regulator but at the 75 mg a day level, the trial subjects experienced no sleepiness. It seems that, like other hormones, melatonin works differently at high doses than it does at low doses.

Hormones are extremely complicated, perplexing and powerful chemicals. There is much more waiting to be discovered about their roles and particularly how they interact with one another. We have seen that as a result of steroid suppression libido calms down but there is evidence that more melatonin can make sex a more pleasurable experience too. Melatonin heightens the effect of our internal endorphins. These are substances that alleviate stress and help to produce sensations of pleasure and relaxed well being. It has even been reported that melatonin, via its stimulation of oxytocin and prolactin, may encourage the physical contact and intimacy that leads to sensual activity. When these hormones were experimentally injected into mice, a dramatic increase in the mouse equivalent of cuddling and hugging took place. Furthermore, when melatonin was added to the evening drinking water of old mice, it was found that not only did they show signs of rejuvenation and increased longevity but also that they engaged in sexual activity again.

Melatonin may help preserve the health of the prostate gland too. This male gland produces the fluid that carries sperm. Nowadays the prostate glands of some fifty-percent of men over the age of fifty become enlarged and this can interfere with both urination and sexual function. That so many men today are inflicted by this ailment suggests that something about our biological system is indeed out of balance.

Proscar, one of the drugs used to treat this condition, works by inhibiting the enzyme 5-alpha-reductase. This enzyme breaks down male hormone into a more potent form that can stimulate the growth of prostate cells. Melatonin also inhibits this enzyme. In fact, when the pineal gland is experimentally removed from mice (resulting in a decline of melatonin) their prostate glands become enlarged but when additional melatonin is given, the gland returns to its normal size.

Research that appeared in 1994 indicates that melatonin plays another crucial role. It has been found that melatonin levels have a significant impact on the structure of microtubules. These extremely small structures are ubiquitous and vital cell components. Although found in cells throughout the body, they occur in particularly high concentrations in the cells of the brain. They are often described as the cell skeleton but their function has been linked to learning, cell organisation and, some have proposed, perception and consciousness too.

Researchers have now linked Altzheimer's disease to declining microtubule function. As melatonin levels decrease with age it would seem likely that a depletion of this versatile chemical is playing a pivotal role in this condition. If all humans are suffering from a deficiency of melatonin, this melatonin/microtubule/consciousness connection could be having an impact on the brains of all of us – particularly as we age.

If in our natural state we had a much more active pineal that pumped more melatonin, many aspects of our biology and particularly our sexuality/fertility would have been different. A lower level of steroid hormone activity and more melatonin would have produced a much more efficient, healthy and a less wasteful reproductive system. We can only wonder what spin-offs this would have had on our consciousness too.

SEX AND DRUGS AND TANTRIC TABOOS ~ψ~

There is a wealth of evidence that humans can still access prolonged transpersonal states of bliss through sexual union. All the major religions, with perhaps the exception of the most patriarchal and fear-inhibited strands of Christianity, have had their sexual mystics and have honoured them. Various Tantric traditions from India to China have used sexual union, in conjunction with awareness of subtle energies within the body, to attain oneness with the 'divine'. The ancient Greeks and Romans also had their cults that combined ritualised sex with magic and often wine too. Sex, drugs and rock and roll has a long history.

From Bacchanalian revels to the rituals of present day underground culture, there are many anecdotal accounts of heightened sexual experiences in combination with particular drug use. (Alcohol, despite perhaps enhancing relaxation and inhibiting fear, in large amounts definitely doesn't enhance sexual experience. Its negative properties are well known and all too commonly experienced.) In Robert Anton Wilson's book 'Sex and Drugs there is the following:

'With pot, sexual intercourse becomes more pleasurable and more relaxed. It makes you a better lover. You feel closer to your partner than you would otherwise. I can feel myself actually fusing with the other person – it is difficult to know even anatomically what part of myself is me and what part is the woman.'

And again:

'And the acid made my consciousness go into the very top eighth of an inch of the head of my penis. That's all I was – just that fragment of flesh pulsating with joy. Then – boom! – I wasn't even that. I was nowhere and yet I was everywhere. Now, that's exactly what the Hindus call Samadhi – union with the All.

Cannabis and tantric practices, we deduce, both help access to the right hemisphere self. Within the bliss of orgasm the small ego sense of the left hemisphere is temporarily suspended. Unfortunately, as noted by no lesser commentator than William Shakespeare, 'normal' orgasm is often nothing more than a 'momentary trick'. D. H. Lawrence called it a 'sneeze in the loins'. Orgasm has routinely descended to a very fleeting and limited experience. By extending orgasm, via practices that often include delay or abstention of ejaculation, profound states can, not only be reached but also dwelt in for lengthy periods. Masters of Tantric yoga are reputed to be able to continue the act of love for seven or eight hours. Aleister Crowley, who dabbled in most aspects of Eastern yoga and Western occultism at some stage in his eventful life, was convinced that sexual yoga was the quickest and easiest way for the average westerner to expand their consciousness.

Dr Richard Alpert, who transformed himself from a clinical psychologist working on LSD to the guru Baba Ram Dass, wrote:

'Tim (Leary) is absolutely right about LSD enhancing sex. Before taking LSD, I never stayed in a state of sexual ecstasy for hours on end, but I have done this under LSD. It heightens all of your senses and it means that you're living the sexual experience totally. Each caress or kiss is timeless.'

Despite Ram Dass' interesting experiences, he concluded that drugs in themselves were not the answer. Such states may be attained more safely and more sustainably using 'spiritual' practices that heighten our consciousness. And these states *are* accessible. Some drugs do have the effect of enhancing the senses but, we may ask, how are they doing this? Some of the effect may be purely chemical hallucination but part may be due to an opening of the gates to a flood of sensory information that we normally filter out. This opening can lead to intense sensations of pleasure that course through the entire body. In such states sensitivity and sensuality is heightened, touch becomes magical and sexual interaction can extend into periods of complete merger with one's partner. There can be a loss of the sense of time and a general feeling of oneness and euphoria in which paradoxically the 'I' that started the process disappears. The body can become the whole universe and the whole universe the body. Something else is accessed in these experiences – something larger or wider than oneself and something that is not restricted by space and time. If the restrictions of space and time stem from the processing capacity of our dominant left hemisphere, from where is this other sense coming?

Perhaps we all know somewhere in the depths of our unconscious that sex offers more than it routinely provides in this day and age of instant gratification. This is the reason why we, particularly in western societies, are so obsessed with sex. We know there is a deeper secret hidden somewhere within our sexuality but most of us are looking for it in the wrong place. The result is a high level of psychological frustration that is externalised to produce a society swamped with sexual images. An uninhibited amount of testosterone coursing through our bloodstream does not help matters. This has not only boosted our libido to perhaps uncomfortable heights but also (as we have seen in hyenas) testosterone can markedly affect levels of aggressive behaviour.

We must also remember that the brain runs on 'drugs'; drugs, despite the connotations of this label, are after all only chemicals. If the brain under our present steroid regime is chronically short of some chemicals, perhaps serotonin and other key neurotransmitters, it is possible that the long history of using DMT mimics in religious/shamanic ritual has been a crude attempt to redress the balance. This cultural use may not have merely been a quest for experiential perceptual effects but for a restoration of health and consciousness. The plants used in these rituals – those that make up South American ayahuasca and Middle Eastern soma – have the same basic biochemical properties. This convergence provides a pointer to the idea that a shortage of DMT may be one missing piece of the consciousness jigsaw.

DMT, THE PINEAL AND SPIRIT ~ψ~

Di-methyl-tryptamine (DMT), despite its complex sounding name, is a structurally simple chemical, derived from tryptophan, an amino acid present in our food. DMT occurs in plants and animals and is part of the normal make up of living things. It has been found in human brain tissue and in our blood and urine, so there is no doubt that we internally produce what is one of the most powerful hallucinogens of all. DMT is particularly abundant in plants of South America, and it is there that, within the cultural traditions of particularly the rainforest tribes, humans have explored its remarkable psychedelic/consciousness expanding properties. In large doses DMT blows open the doors of our normal perceptions to allow access to worlds beyond our imagination. Fantastic visions, out of body travel, near-death experiences, predictions of the future, contact with the dead and 'alien' presences, are all part of the DMT realm. Some of these experiences sometimes occur spontaneously, without resort to any extraneous chemicals. There are even cases in which individuals are totally convinced they have been taken to another world by aliens. If drug-induced and naturally occurring mental conditions appear to overlap, it certainly hints at some natural function for this endogenous brain chemical.

If one chemical can do all this, many questions arise about the nature of our normal consciousness. For instance, is our usual perception dependent on a delicate balance of such molecules? In the early 1950s, the discovery of the related tryptomines, LSD and serotonin, raised such questions, and rocked the foundations of the comfortable domain of psychiatry. In 1955, Hungarian chemist and psychiatrist, Stephen Szára, having been frustrated by unsuccessful attempts to procure LSD and mescaline for a research project, synthesised some DMT in his Budapest Laboratory. He then tried it out himself, at first, by eating it. This had no effect; it was later discovered that there is a mechanism in the gut that breaks down DMT. (The secret of the South American tribal brews, like ayahuasca, is that they include chemicals that inhibit this mechanism.) Szára then injected DMT and its effects became very evident indeed. He wrote:

'The hallucinations consisted of moving, brilliantly coloured oriental motifs, and later I saw wonderful scenes altering very rapidly. The faces of people seemed to be masked. My emotional state was elevated sometimes up to euphoria.'

Szára then co-opted thirty volunteers who were all given the full 'psychedelic' dose. Their accounts are extremely interesting, particularly those that speak of spiritual experience. Here are some fragments: 'The whole room is filled with spirits.'… 'I feel exactly as if I were flying' ... 'Everything has a spiritual tinge but is so real.' ... 'In front of me are two, quiet sunlit Gods…I think they are welcoming me into this new world.' … 'I am finally at home.'

With the explosion of interest in hallucinogenic drugs in the 1960s and the subsequent authoritarian backlash, legitimate research into LSD and DMT was halted for a generation. This was a shame for some of the research was very promising. For example, consciousness researcher Willis Harman found that LSD had a very positive effect on creativity. Now, in this more relaxed era, a few studies are being undertaken again. Scientists at the John Hopkins Medical Institution have recently found that even in carefully controlled conditions the 'sacred mushroom' chemical, psilopsybin (another tryptomine), induced mystical experiences that led to positive change in their subjects' outlook. One third (out of 60) said the experience was the single most spiritually significant of their lives and more than two thirds rated it among their top five most meaningful.

However, the most notable work so far has been Rick Strassman's investigation into what he calls 'the spirit molecule'. Such research is deemed important, not only for the light it shines on the nature of the human consciousness system, but also for the potential of these chemicals to help resolve deep psychological problems.

After a lengthy battle with the relevant U.S. State departments to gain the go ahead, and further difficulties in procuring the highest grade of DMT, Strassman finally administered the drug to a carefully chosen group of volunteers. Most of these found the high dose of DMT exciting, euphoric and extraordinary pleasurable. Sometimes this elation related to the unfolding visions themselves and sometimes it arose from the revelatory insights that were gained during the sessions. For some, however, the experience was extremely frightening and this could have been, in part at least, to the almost complete loss of control the participants felt.

Intriguingly, some of those that managed to go beyond this loss of control barrier, noted an 'outside' intelligence or force directing their minds, and a few of these believed they had contact with beings. Several volunteers experienced 'abduction by aliens' and interactions with them. Despite the bizarre nature of these scenarios, they felt very real. One volunteer reported that 'it felt more real than real'. Though the high doses of the hallucinogen were without doubt creating all sorts of perceptual peculiarities, these incidents are reminiscent of the split consciousness effects within schizophrenics. Were the alien voices and visions arising from the part of the mind beyond the ego mind? They had to come from somewhere! The unlikely alternative is that they really were coming from an external source and the DMT was facilitating some sort of access. This is an idea that Strassman seriously considers. His findings suggested that DMT provides regular, repeated and reliable access to 'other' realities and these could be thought of as something like TV channels. He then goes on to ask whether these other channels of existence are always present, 'transmitting' all the time, but under normal circumstances are not perceptible. He speculates that our natural (chemical) balance keeps us tuned to 'Channel Normal' but, when subjected to a flood of DMT, our minds open to these other planes.

Does DMT remove a filter that then allows access to different dimensions? It would take a lot to convince us that there really were aliens out (or even in) there. But it is possible that there could there be a link here to a filter imposed by the left hemisphere self. We have seen in the last chapter how the brain, under certain conditions (as demonstrated by autistic savants) can take in vastly more detail than it does when working 'normally'. If DMT was allowing the brain to open to a flood of sensory data as well as distorting that data, it could explain some of the reported results. Much more research needs to be done in this area, but it may be significant that the 'aliens' tended to communicate using a language of universal visual symbols rather than sounds and words; a means of communication more in tune with the right-hemisphere self.

Whenever possible Strassman tried to recruit volunteers who had already had some experience of hallucinogens. It appeared that those who were more familiar with the effects of mind-altering drugs were less prone to fear and were less likely to project such things as alien encounters into their experience. Could it be that the left hemisphere, which we know tries its best to make up stories to fit experiences it cannot cope with, was grasping at ideas of aliens to try to get a handle on the flood of unusual sensations? Certainly this would be the simplest explanation for these bizarre encounters.

One further effect that was experienced by many of the DMT volunteers was a loss of normal time perception. Most believed their entire session had only taken a few minutes. Many felt that, at the highest point of their DMT experience, they entered a timeless zone, but within that zone an enormous amount happened. We can also note here that, according to our speculation, the right hemisphere self functions within the timeless present, whereas the left functions within a constructed time scale of past events and anticipated futures. All our worries and fears occur within these projected places, thus by being forced to step outside these ephemeral time zones, insight can be gained into the nature of our personal problems. Strassman found that the euphoria brought on by DMT helped volunteers to look at their lives and conflicts.

While the most bizarre and extreme effects came from high doses of DMT, Strassman experimented with smaller ones too. The lowest category of dose was experienced as pleasant, and almost all the volunteers said they felt like smiling or laughing after receiving it. If we can accept that our usual mental state of something bordering on worry and fear is an imbalance and not 'normal', it is possible that this imbalance is due to a chronic shortage of this natural brain chemical. Strassman thought there was something 'peerless' about DMT despite its overt chemical and hallucinogenic similarities to LSD. One of its differences though is the speed of its action – the hallucinogenic effect is felt within a few seconds of it being intravenously administered, and the entire 'trip' only lasts some ten to fifteen minutes. This points to something that is unique about DMT. It appears that our brains have some mechanism that rapidly 'consumes' this chemical.

DMT is the simplest of the tryptamine psychedelics and the smallest. Its molecular size is only slightly greater than glucose. The brain, being a highly sensitive organ, possesses a nearly impenetrable shield, the blood-brain barrier, which prevents unwelcome chemicals leaving the blood and entering brain tissue. Even complex carbohydrates and fats that other tissues use for energy are kept out. The brain uses only the purest form of fuel – glucose. However a few essential molecules, like the amino acids needed for the construction of brain proteins, are actively transported across the brain-blood barrier. Twenty-five years ago, Japanese scientists made the startling discovery that DMT is one of the select number of chemicals that are actively taken into the brain. Once in the body or brain, certain enzymes break it down within seconds. These enzymes are none other than the monoamine oxidases (MAO) that we have come across before – the very enzymes that are inhibited by the bioflavonoids found in fruit. Strassman points out that the brain is acting almost as if DMT is a 'brain food' like glucose. Both are part of a rapid turnover system. All this suggests that DMT is a highly significant part of our functional make-up but what is it actually doing?

One of Strassman's speculative conclusions is that a certain level of DMT is needed by the brain to keep it on the perceptual straight and narrow – that is, it acts as a 'reality thermostat' keeping us within a narrow band of experience. Too much of the chemical and all manner of unusual visions and feelings appear on our 'mind screens'. Too little and our view of the world dims and flattens. These latter effects are just what normal volunteers feel when they are experimentally given anti-psychotic drugs. It is possible that these 'medicines' cause such depressive symptoms by blocking the production or action of endogenously produced DMT. If this were so, who is to say the levels of DMT that we produce today are correct for optimal human performance? If our brains have suffered some degeneration over the last however many thousand years perhaps our DMT levels are a little low.

If this is so then the discovery that LSD does not work in the left hemisphere (Serafetinides – see Chapter One) takes on a new and greatly heightened significance. If LSD, for whatever reason, has no perceptual effect on the left hemisphere, does the structurally very similar DMT? Further research into this area would be extremely enlightening. It is just possible that, if the left hemisphere is damaged and cannot respond to its optimal quota of DMT, this may be affecting our perception and indeed our sense of who we are. As all our individual and collective problems stem from our selfish, ego-based, disconnected sense of self, all this has profound implications.

Though a diagnostic experiment has yet to be done, circumstantial evidence points to the pineal gland as the site of DMT production. Most spontaneous experiences of alien abductions occur in the early hours of the morning – just at the time the pineal is at its most active. The pineal contains the highest concentrations of serotonin in the body. This chemical is the raw material for melatonin and in all probability DMT too. The pineal has the ability to turn serotonin into tryptamine and it also contains high levels of methyltransferases – the enzymes that attach methyl groups to other molecules. These enzymes only have to do their job twice to construct di-methyl-tryptamine (DMT). As we have noted in earlier chapters, the pineal also makes beta-carbolines, and it is these compounds that inhibit the breakdown of serotonin and DMT too by blocking the action of the body's MAO. Thus the pineal gland may not only produce DMT but also the chemicals that prolong its activity.

But how could DMT production be activated? We know that melatonin synthesis in the pineal is 'turned on' by the neurotransmitters, noradrenaline and adrenaline, which are released by nerve cells that almost touch the gland. We also know that the adrenal glands produce these two neurotransmitters and release them into the bloodstream in response to stress. Thus stressful conditions could potentially upset pineal activity, particularly as the pineal exists outside the brain-blood barrier and so should be responsive to blood borne chemicals. However the pineal has its own security system that protects it from such interference. This mechanism is very efficient: it makes it difficult for melatonin to be produced during the day but even so the system can be overridden. We have already discussed how meditation and cannabis use can initiate melatonin production and boost levels in the blood to a marked degree. Could parallel mechanisms initiate DMT production?

For the production of enough DMT to precipitate extraordinary visions, the security system around the pineal would have to be overcome as well as the activity of the methyltransferases boosted and MAOs inhibited. Some research suggests that, if the body/mind is subject to enough stress, it is possible to breach the pineal gland's defence shield. It has been found that DMT levels rise in animals exposed to stress, and, in psychotic individuals, stress intensifies hallucinations and delusions. If stress then precipitates excessive

DMT production, it could help explain such bizarre perceptions as alien abductions. That these experiences often coincide with personal crisis, loss or trauma further strengthens this case.

To add a further twist to this fascinating story, it is possible that the odd experiences and weird visions are actually distorted interpretations arising from the limited perceptual abilities of the left hemisphere. If the dominant hemisphere is presented with an experience beyond its computational abilities, its attempts to make sense of it may result in a distorted fear-based picture.

To tie this all together then, we can tentatively conclude that DMT is produced in the pineal, used in the brain and that overproduction, brought on by stress, can lead to strange experiences. DMT affects how we feel. Low doses have been found to bring on feelings of laughter and happiness, and a shortage of the chemical has been associated with depressed states. DMT may be acting in the brain therefore as some sort of regulator of consciousness. It is also possible that the chemical does not work in the left hemisphere as it does in the right. If this were indeed the case, particularly as our left hemisphere sense of self is dominant, the implications would be enormous.

Before leaving this discussion on DMT we should touch on the subject of 'spirit'. The pineal has, in many cultures, been associated with spiritual experience. It has been called the 'third eye' and the 'seat of the soul'. It is associated with internal light and spiritual illumination. Such spiritual experiences that include visions of angels, hearing heavenly sounds, a sense of timelessness and near-death experiences occur within all religious traditions. They are also characteristic of a fully psychedelic DMT trip. So if stress can cause the pineal to produce DMT can its opposite, meditation, do likewise? Rick Strassman believes it can.

Meditation (as well as prayer, chanting, visualisations) can effect the pineal and brain activity by shifting the balance between left and right hemispheres. Studies show that in meditation, brain wave patterns are slower and better organised than those produced during our usual states of awareness. These brain wave patterns can lead to a deep state of bliss that we can feel throughout the body. Within the brain, Strassman speculates, these patterns may induce a resonance that would effect all our structures. It is just possible that such a resonance within the pineal could weaken the barriers to DMT formation resulting in a surge of the chemical that could become stronger as the meditation became deeper. The highly blissful states that can be attained by meditation are certainly similar in nature to the euphoric ones experienced by Strassman's volunteers. Furthermore, in the light of two of our previous sections on fertility and sex, we can also make the connection between these highly euphoric states and those attained during the deepest, most loving orgasms. We suspect that DMT is playing its part here too.

Ecstasy and Empathy

Another closely related tryptamine that made a big impact, particularly during the 1990s, is 3,4-Methylenedioxy-N-methamphetamine (MDMA). This drug for a short while became a social phenomenon – it was used recreationally by millions of young people from all backgrounds, and it changed their behaviour. Not only did it produce, like DMT, a feel-good factor but also encouraged hugging, closeness and bonding. The drug ended up with the street name 'Ecstasy' but one of its first distributors in Los Angeles preferred the name 'Empathy', as this was more descriptive of the drug's true nature.

Now, a decade later, researchers are looking into both the psychotherapeutic properties of MDMA and its brief sociological impact. At its height, the influence of the drug reached beyond the typical night club and trance dance scenes – in Northern Ireland, Protestant and Catholic teenagers dropped their conditioned hatred and spontaneously started to hug one another. Even arrests for football violence dropped by 22%. Many fans said this was because Ecstasy had replaced beer as the drug of choice on the terraces. Rival gangs of supporters met in clubs, took the pills and, within minutes, a lifetime of enmity disappeared. Some documentary research has even raised the idea that MDMA played a part in instigating the famous First World War football games between German and English soldiers. Apparently the drug was trialed as an appetite suppressant but quickly dropped by the military due to unwanted side effects.

Although playing around with potent neurochemicals is always open to dangers, especially when used in uncontrolled conditions, Ecstasy in itself proved remarkably safe. The well-publicised casualties of the drug were more to do with ignorance of use, inappropriate drug combinations and impure or adulterated supplies than with the chemical itself. In fact pure MDMA, and yet another tryptamine (ketamine), are now being investigated as potential psychotherapeutic tools. Research into their properties has discovered that they have the ability to reset the brain in some way (by updating the left hemisphere's conditioned reality perhaps). Ketamine has been found to be particularly effective in 'rebooting' the brain in instances when clinically unhelpful responses, particularly depression, are resistant to change. Further research into the activity of MDMA has found that it causes a brain surge of oxytocin – the hormone that helps to bond couples, and mothers to babies – a factor no doubt responsible for the overt changes in behaviour that is the hallmark of this drug.

The fact that MDMA has the capacity to instantly and, in many cases, permanently transform the conditioning and cultural beliefs responsible for enmity, aggression and violence deserves thorough consideration. It is also significant that these types of drugs can relieve depression and induce joy. Whilst we are certainly not suggesting that humans are deficient in MDMA, it is likely that we are clinically deficient in key neurochemicals that have some similar properties. That there is some deep neurochemical problem is undeniable – 54 million people have taken the prescribed brain drug Prozac since its launch in 1986 and depression affects over 120 million people world-wide. 850,000 depressed people kill themselves every year and, according to the World Health Organisation, depression is the fourth largest contributor to the global burden of disease.

Mystic and religious experience almost certainly involves a change in brain chemistry. Indeed, loss of connection with 'God' may be partly due to a biochemical deficiency once essential for 'normal' brain function. Hebrew scholar, Carlos Suares, has commented that:

'We must remember that the only instrument of investigation we possess is our mind. If we do not completely understand how our mind works, this instrument will twist and disfigure whatever 'reality' we may discover. The quality and condition of the telescope govern the observation resulting from its use. If there is dust on our lens, we see dark spots in the heavens.'

Ironically, over the last few thousand years, the switch from experiential to conceptual religions (itself a reflection of a shift in cerebral dominance) has increased the 'dust on our lenses'. Because the left hemisphere's 'sense of self' fears right hemisphere experience, such patriarchal religions ban, suppress, or at least frown upon the use of neuro-chemical analogues, yet, at the same time, remind us of our current state of disconnection. Despite heaven being described in terms of rapture, any form of pleasure is proscribed – sex, music, dance and singing all require the engagement of the right hemisphere and are regarded with varying degrees of hostility. Female circumcision, still practised to this day is a good example of this. Extreme factions of many religions would have undoubtedly tried to outlaw sex if it had not been necessary for procreation. It was not always so – the Gnostic Carpocratian school (that predates the authoritarian Christianity that became dominant in the Third Century) taught that good and bad conducts are only matters of human opinion, not divine decree. Students were encouraged to enjoy life, including sex. Epiphanes, the son of Carpocrates, writes:

'God created the delights of love equally for all humankind. But men have repudiated the very thing that is the source of their existence.'

Such Gnostics, preceding Ecstasy users by 2000 years, not only saw sexuality as a celebration of the union of God and Goddess but also practised sacramental nudity and ritual intercourse as a way to reconnect to the 'mystery'. Divine rapture and religious ecstasy, however approached, is facilitated by unimpeded right hemisphere function – something alluded to by modern day goddess, Kylie Minogue, in the Karen Poole and Johnny Douglas song 'Red Blooded Women'. The clever lyrics 'You will never get to heaven, if you are scared of getting high' are a concise and poetic way of saying that to access blissful states of mind, the anxiety felt by the left hemisphere self needs to be overcome.

THE IMMUNE SYSTEM ~ψ~

The thymus gland is the most central and essential organ of the body's immune system. It lies in the upper chest and in infants and children it is relatively large, commonly extending from the base of the neck to the area of the upper heart. The thymus reaches its maximum size and weight (about 40 grams) just before puberty. After this significant life change, it gradually diminishes in size and by the age of 50 it will usually weigh less than 12 grams. It is believed that the decreased size and secretory abilities of the thymus may make the elderly more susceptible to disease.

The thymus produces several hormones that are important to the development and maintenance of our immunological defences. These thymic hormones (thymosins) promote the development and maturation of the white blood cells (lymphocytes) which are the 'sharp end' of our immune response. Approximately 80% of circulating lymphocytes are 'T-cells' and these are dependent on the thymus gland. There are many different types of T-cells and all are important in maintaining the health of the body. Some attack foreign cells or cells infected with viruses. Some stimulate regional inflammation and local defences in injured tissue. Some are 'helper' T-cells that activate both T-cells and B-cells (those made in bone marrow), and some are 'suppresser' T-cells that act with the others to regulate and fine tune immune response.

The thymus then is of crucial importance but strangely at puberty, in response to increasing levels of steroid hormones, it starts to shrink. As it does, its internal structure becomes so 'disorganised' that many researchers believed that it ceased to function altogether. However, Richard Boyd and Jayne Sutherland of Monash Medical School in Melbourne have recently found that, in mice at least, the thymus does continue to produce T-cells but only at about one tenth of the rate it does in a young animal. Furthermore, when they physically castrated mice they found that the thymus regained its youthful appearance within four weeks and that the number of T-cells it produced increased to near pre-pubertal levels.

In a related study, Richard Koup and colleagues at the University of Texas Southwestern Medical Center in Dallas found that, as in mice, the human gland continues to function after puberty but at a similarly reduced level. Koup also found that the thymus produced increased levels of T-cells in HIV patients receiving aggressive treatment with combinations of various AIDS drugs. This strongly suggests that boosting the gland's function may help combat AIDs and possibly other invasive diseases too. This research coupled with the Australian work points to extremely powerful therapeutic tool. If drugs could be given to suppress the action of sex steroids, the thymus may regain some of its size and function. If it starts to pump out more T-cells, the immune system will be enhanced and viruses that were thought to be too much for the human immune system could be dealt with not only efficiently but also in a much less toxic way.

This could have far-reaching medical consequences for us today but, if in our past the activity of these steroids were suppressed, a thymus that functioned fully, perhaps even into old age, may have been the norm. This would have had attendant effects on health and longevity. The unique circumstances within our evolution that we have outlined in Chapter Two would not have only produced a big brain but also a much more efficient immune system. Could we today then be running a system with only a fraction of its potential? As it is now, our immune system builds but then is suppressed at puberty. It is still pretty amazing (as all human systems are) but possibly it is only a shadow of its former self.

If we all have the potential of a much more powerful immune system, it is conceivable that, from time to time under unusual circumstances, it may be stimulated to come on stream. Such an alternative pathway could explain aspects of 'spiritual' healing, spontaneous remissions and even old tales of a golden age when everyone was reputedly long lived and bounding with health. There is something in all this that is worthy of further investigation.

The overall picture is simple and wouldn't be difficult to conclusively prove. We already know that when steroid hormones are suppressed the thymus reactivates and can act powerfully. This is regarded as an anomaly but, if in fact the anomalous feature is our overactive steroid hormones, the whole way we look at our immune system radically shifts. The activity of steroid hormones that we regard as normal today significantly suppresses our immune function. If the levels of these steroids were experimentally reduced or inhibited by increased levels of melatonin, beta-carbolines and nutritional chemicals, our immune system should be enhanced. This, we surmise, would have been the immunological scenario 'before the fall'. As these steroids are the same ones that Geschwind and others suggest have a deleterious effect on particularly the left cerebral cortex of the brain, the picture could be broader still. Just by inhibiting these steroids everything could start to function very differently.

When the body/mind is under stress, cortico-steroids are produced in the adrenal glands and the brain. There is a recognised link between stress and immune function. When we are stressed our immune function doesn't work so well – we are more prone to catch colds because our resistance is lower. Stress thus suppresses immune function via the cortico-steroids. These are structurally very similar to our steroid sex hormones that also suppress our immune function. (That steroids do suppress the immune system is widely acknowledged for they are administered to transplant patients to help suppress the body's immune rejection response to the alien tissue.)   Stress and fear are very damaging.

As noted before, steroid hormones can cause developmental damage to particularly the left cerebral cortex. This may have some affect on immune function because, due to this damage, our dominant left-hemisphere self now suffers from a perpetual background state of fear that is of course stressful. And this stress will result in higher levels of cortico-steroids being produced. Thus within our less than perfectly functioning neural systems, there are loops of damage that interact with one another. If we could be released from all this internal warfare what would emerge?

When we meditate we reduce our internal levels of stress. We know too that meditation results in more melatonin being produced, which will suppress steroids and in turn will enhance immune function. Meditation also affects consciousness. Reaching a state of quietness by calming the internal verbal dialogue suggests shifting the balance away from left hemisphere dominance. Less stress means less cortico-steroid activity, which means less of a negative impact on the immune system. All these elements will improve health and well being.

There are further clues that point to the possibility of latent and enhanced physiological (and consciousness) functions within some humans. We have touched on these before. People with multiple personality disorder, for example, can heal at a faster rate than is normal. Perhaps they are not so locked into a one-sided neuroendocrine control loop – they may benefit from more right hemisphere involvement. If the right neo-cortex runs a more efficient neuroendocrine system than the left, it may explain other anomalous events in the healing arena too, such as the link between recovery from illness and feverish visions. And as the neo-cortex fine-tunes many other functions, shifting dominance may result in a body run in a much more efficient manner. Even our digestive system could be affected by such a change.

A SECOND BRAIN ~ψ~

Research by Dr Michael Gershon has shed new light on the murky world of animal digestion that is of significance to our own story. Almost single-handedly, he has elevated what is generally regarded as a mundane process to the dizzy heights of neuroscience. He has shown that the gut is definitely more than a tube that processes food. It contains a highly complex neural system and displays a great degree of autonomy.

Gershon's major breakthrough came about when he demonstrated that serotonin was extremely active in the gut. In fact, it is now known that the gut uses 95% of the body's serotonin. Since his initial discoveries, every neurotransmitter known in the brain has been found in the gut too. Furthermore, the human gut has a complex self-contained nervous system containing more nerve cells than the spinal cord, and indeed more neurones than all the rest of the peripheral nervous system. There are over 100 million nerve cells in the human small intestine alone. Structurally and neuro-chemically, the enteric nervous system (ENS) is a brain unto itself. Within those yards of tubing lies a complex micro circuitry driven by more neurotransmitters and neuromodulators than can be found anywhere else in the peripheral nervous system.

Though connected to the brain by the vagus nerve, it has been found that the gut has a great degree of autonomy. In 1899, physiologists studying dogs found that, unlike any other reflex, the continuous push of material through the digestive system (now called the peristaltic reflex) continued when nerves linking the brain to the intestines were cut. The vagus nerve only directly connects with a comparatively tiny number of gut cells and while it has overall control, the brain does not instruct the gut *how* to carry out specific tasks. It is strictly an inside job, and one that the gut is marvellously capable of performing. In addition to propulsion, the ENS bears primary responsibility for self-cleaning, regulating the luminal environment, working with the immune system to defend the bowel, and modifying the rate of proliferation and growth of mucosal cells.

The gut is a major 'immune organ' too, containing more immune cells than the rest of the body combined. The enteric nervous system interacts intimately with the immune system, and can affect mood and behaviour by signalling the central nervous system. Indeed, the vagus nerve mostly carries information from the enteric nervous system to the brain not vice versa. For every one message sent by the brain to the gut, about nine are sent in the other direction. Feelings of fullness, nausea, the urge to vomit and abdominal pain are all the gut's way of warning the brain of danger from ingested food or infectious pathogens. And recent research has found that stimulating the vagus nerve can have antidepressant and learning-enhancing effects. It is even now known that the vagus nerve provides an alternative neural pathway for the orgasmic response. Gut feelings are genuinely more than just a metaphor.

Melatonin may play its part in this. Though the pineal is responsible for most of the melatonin that circulates around the body, it is now known that synthesis of melatonin can occur locally in cells throughout the body. The gastrointestinal tract is a major source of extra-pineal melatonin. Here melatonin protects the gut from ulceration by its antioxidant action, by stimulating the immune system and by fostering micro-circulation and epithelial regeneration. We also know that melatonin monitors our mood, thus it is possible that gastrointestinally produced melatonin may play some part in our feelings of well being.

There are clinical implications to all this too. Because the neurotransmitters and neuromodulators present in the brain are nearly always present in the bowel as well, drugs designed to act on serotonin metabolism are likely to have enteric effects. About 25% of patients taking antidepressant medicines report some initial nausea or diarrhoea.

Mood-altering drugs like Prozac, acting simultaneously on both brain and gut systems, may have even more effect on the bowel than on the brain, because serotonin predominates in the bowel and the drug moves through the digestive system before reaching the brain. Fortunately, in most people, the bowel quickly develops tolerance to these drugs, and gastrointestinal side effects usually subside within a few days of the starting treatment.

From our perspective these discoveries have great significance. The digestive and assimilation system is effectively a tube of complex neural tissues, as responsive and delicate as our cerebral brain and running on the same biochemistry. We are dependent on our gut for the assimilation of our biochemical raw materials. If the balance of steroid hormones and hence neurotransmitters has been upset, this will affect how the very delicate gut mechanism works. If the gut is not at optimum function then the assimilation and absorption of essential nutrients will be impaired.

The diet of our forest ancestors was a rich mixture of leaves and fruit. Each element of this diet contained thousands of unique chemicals. Whole groups of these chemicals are known to affect neural biochemistry and some in similar ways to anti-depressant drugs. Both elevate the activity of neurotransmitters. For perhaps millions of years a diet of several kilos of plant material, loaded with thousands of chemicals, were eaten every day. These chemicals would have become an integral part of the operating biochemistry of the 'second brain'. Indeed the gut (like the cerebral brain) must have developed and adapted in response to this high-powered fuel. Its operation may have become partially dependent on the continued flow of chemicals, such as MAO inhibitors, absorbed from fruit passing through it. The loss of this forest biochemistry could have altered, perhaps significantly, the functioning of the digestive system.

These same chemicals have had a similar effect on our brains. As the neural systems of brain and gut are directly connected (via the vagus nerve) and this two-way traffic of information influences the function of both, it is possible that the interaction and resulting operation is now less than optimal. In the distant past, there may have been a more unified and powerful connection between the two systems, and both would have run more efficiently as a result. It is even tenable that a specialist fruit diet, coupled with an increasing synergy between the two neural systems, initiated a new and powerfully efficient layer of function. The vagus nerve rather than being a mere conduit for the flow of information could have acted more like the corpus callosum. The second brain may have been, in effect, something like a third hemisphere.

If we could restore the missing biochemistry and shift cerebral balance, a different level of digestion and assimilation could emerge. Just eating more figs though would not be enough. To fully engage the system, appropriate biochemistry plus a significant shift in cerebral dominance would be needed. A combination of an enhanced neuroendocrine function and enhanced communication between the brain and the gut, plus the direct effects of diet could be enough to engage a level of function well outside current norms. And more efficient assimilation could help restore all our body/mind systems.

Gershon and others point out that a rise in serotonin in the gut can bring on diarrhoea and increasing it further can create constipation. It is interesting to note the comparison between this observation and the effects of prolonged sleep deprivation (increasing a shift to right hemisphere control) on bowel function. On one of his sleepless trials, A.W. had a brief experience of diarrhoea that resulted in the expulsion of partially undigested food. This was followed on day five by something that, though not constipation, was an unusually small, dense bowel movement. The gut area itself changed shape. It became somewhat rounded, sensitive but free from any tension. It felt like having an active rather than a passive gut. Could this have been the first stage of a transition to a gut with an enhanced function?

Michael Gershon's fascinating research has taken no account of human origins in regard to forest biochemistry. What he has done is to show that the gut is an extremely sensitive and delicate system. In our evolutionary past, this system would have been flooded with a complex and unique biochemistry. In a personal communication, he has commented that if the primordial diet were rich in monoamine oxidase inhibitors, it would have profoundly affected the evolution of the gut. He agrees that, at least in theory, a diet rich in fruit would have had some sort of effect on this process.

Our neuroendocrine, immune response, and assimilation systems are all closely linked – they are all interdependent parts of a unique system that was once fuelled and built from tropical forest biochemistry. The effects of the loss of this supporting biochemistry on each part of the system have been significant. But, as optimum function of one part of the system depends on optimum function of the rest then any loss in any part would create a domino effect. Any attempt to redress the balance would need a combined and simultaneous consideration of the whole system, not its component parts.

SUPERMAN/SUPERWOMAN ~ψ~

A recent television documentary covered the story of a man involved in a shipping accident who fell into the sea. He should have died quickly from the low temperatures. However, he not only survived but also rescued a few other floundering people along the way. When he was finally picked up, and was safe in a helicopter, he described a mental state of surrender that allowed something more functional and powerful to take over. There are many similar stories of seemingly miraculous feats of endurance, strength and heroic valour that are remembered for their effortless dream-like quality that often incorporate mental states of clarity and fearlessness.

The annals of sport are also full of extraordinary performances that seem to go beyond the normal range of human accomplishment. Rhea White, co-author (with Michael Murphy) of 'In The Zone', has collected accounts of over 4,500 of these, many of which appear to have a quasi-spiritual aspect. For instance, John Walker's description of his win in the 1500 metres at the 1976 Olympics suggests that an altered mind state was a crucial factor:

'... when I hit the front I got a flash of compelling certainty. I didn't look over my shoulder, but I sensed someone coming up on me fast ...I was already at full stretch. But I went into a sort of mental overdrive, and my subconscious mind took over completely – I've experienced it in races before and I can't explain it. I burned Wohlhuter off and went to the tape with my hands over my head.'

Another runner's experience further illustrates this ability to access a part of ourselves that is normally locked away – Dr George Sheehan relates:

'The first 30 minutes is for my body. During that half-hour I take joy in my physical ability, the endurance and power of my running. I find it a time when I feel myself competent and in control of my body, when I can think about my problems and plan my day-to-day world. In many ways, that 30 minutes is all ego, all the self. It has to do with me the individual.
What lies beyond this fitness of muscle? I can only answer for myself. The next 30 minutes is for my soul. In it, I come upon the third wind (not the second wind, which is physiological). And then, I see myself not as an individual but as part of the universe. In it, I can happen upon anything I ever have read or saw or experienced. Every fact and instinct and emotion is unlocked and made available to me through some mysterious operation in the brain.'

It is particularly fascinating, in the light of our previous investigation into memory, that Dr Sheehan found a way to access what appears to be total recall via this route of physical activity. His experience is not unique. After a golf tournament, Jack Nicklaus can remember every stroke he played, and similarly, American Ball player Whitey Ford is said to be able to recall every pitch from over 3,170 innings. Could the exceptional states of mind accessed at these times determine both the quality of performance and the vividness of the recollection?

Stillness, peace and a liberated detachment from our ordinary state of mind are qualities usually associated with meditation and deep prayer. But such states can take over too in the midst of extremely focused activity: when he broke the four-minute mile, Roger Bannister said he felt a 'complete detachment; there was no pain, only a great unity of movement and aim. The world seemed to stand still, or did not exist.' David Hemery, who set a world record at the 1968 Olympics in the 400 metre hurdles, wrote that his mind and his body worked almost as one. His limbs reacted as his mind was thinking 'total control', which resulted in a state of 'absolute freedom'. And the tennis player, Billie Jean King, has written about playing the perfect shot:

'I can almost feel it coming. It usually happens on one of those perfect days when everything is just right, when the crowd is large and enthusiastic and my concentration is so perfect it almost seems as though I am able to transport myself, beyond the turmoil on the court, to some place of total peace and calm.'

Another record breaker, speed skier Steve McKinney, recalls that at such peak times it was like 'riding the substance of dreams, a magic carpet of air, into which power was sensuously entwined'. And marathon runner, Ian Thompson, has only to think of putting on his running shoes to be filled with a kinaesthetic pleasure of floating. Ecstasy is a remarkably common experience felt by many sportsmen, dancers and even musicians.

Some barrier needs to be broken through to reach these levels of high performance and effortless being. The process appears to involve (yet again) moving beyond the left hemisphere's sense of self. This view is strengthened by the experience of Mike Spino who, after an extraordinarily fast and effortless run, found that:

'When the run was over conversation was impossible, because for a while I did not know who I was. Was I the one who had been running or the ordinary Mike Spino? I sat down by the roadway and wept. Here I was, having run the entire six miles on a muddy roadside at a four-and-a-half minute pace, which was close to the national record, and I was having a crisis deciding who I was.'

It is theoretically possible that these breaks from the normal ego sense of self entails a shift into a more primitive level of brain function – an animal level in which physical movement is enhanced. But, as such fluid activity is often accompanied by what we could regard as the 'highest' mystical or transpersonal states, something much more interesting could be happening; a shift to a second self with a very different sense of identity and enhanced abilities. Whatever it is that is happening, one thing that seems undeniable is that the left hemisphere's linear sense of self appears to have serious problems with these experiences. As highlighted in the last example – all it can do is sit down and cry.

Sometimes after the event, when a more usual consciousness is resumed, these altered states can even be felt as frightening. Perhaps it would be more accurate to say that something within feels fear when it re-establishes its own centre of self. When Bob Beamon broke the world long jump record, by the astounding measure of nearly two feet, he could not believe it. Waves of nausea rolled over him, his heart started to pound and he saw stars in front of his eyes. Athletes can find that such extraordinary, one off performances go way beyond what the logical mind can accept and understand. A conflict arises that is felt, in extreme cases, as terror. How can we explain this? It seems that at these times the ordinary linear mind, which is constantly attempting to maintain a veneer of control, is having its authority threatened by the undeniable display of second self function. Its response is its habitual one of confusion and cover up, overlain with fear.

Murphy has also pointed out that 'many athletes have trouble recapturing peak moments, because they have trouble incorporating the meaning of these experiences into the rest of their lives'. We would suggest that it is just their ordinary mind that is having the trouble here, and it is to be expected. Such peak performances, like dreaming and hypnosis, do not happen within the arena of that self. It seems sportsmen are reaching these exalted places almost by accident. The left hemisphere sense of self, and its control, is being overridden without it really noticing, but, when it does, it reacts with fear and confusion. There is an old Indian story that tells of a thief who pretended to be a yogi in order to escape retribution. His dedication to the pretence was so good he became enlightened. The parallels with athletes are obvious. They practise, focus and concentrate for the sake of their craft, and occasionally catch glimpses of spiritual freedom through the discipline. The meditation master tries to overcome the 'monkey mind' – a term given to the unruly ego-based mental system – more directly.

Being truly 'in the present' can be the bridge to this goal. The habitual mind tries to deny or resist the present because it doesn't seem able to function or remain in control without time. Time is always the past and the future. This mental mechanism therefore perceives the 'timeless now' as threatening to its construction of self. Also, as sporting activity is always happening in the 'now zone', having a controller acting from a mental position of past experience or anticipated future is not conducive to optimum performance. This part of the mind is not fast enough to cope with the split second decisions and responses that need to be made during the heat of activity. It also seems that when we access the part of us that functions within the 'now zone', the ego-based habitual mind is no longer present. This feels very good and is why sport (and meditation) can be addictive. The Formula One Motor Racing hero, Jochen Rindt, once said that, when driving: 'you forget about the whole world and you just … are part of the car and the track. It's a very special feeling. You are completely out of this world. There is nothing like it'. Golfer Arnold Palmer noted too that concentrating on the shot at hand induced a heightened sense of presence and renewal that could endure through an entire round.

It appears that a second system can begin to operate once it is released from the ordinary mind. E.J. Harrison tells of a demonstration, given by Matsuura, a high-ranking instructor at the Kodokan School of Judo, that clearly demonstrates this point:

'Sitting on his knees with his back to me and his hands together, he made his mind blank of any conscious thought. The idea was that I was to remain behind him for as long a time as I desired. Then with all the speed and power I could muster, I was to grab him by the throat and pull him over backwards. I sweated it out for maybe two or three minutes without making a move. Then I put all the power and speed I could into the effort. My next step was to get up from my back where I landed in front of him. His explanation was that the action was not conscious, but rather sprang from the seat of reflex control, the tanden*, or second brain.'*

Accessing what some people assume to be an 'unconscious reflex' is a secret that lies at the heart of enhanced sporting performance. In Zen and the Art of Archery, Eugen Herrigel was taught that the shot would only go smoothly when it took the archer himself by surprise. He was taught not to open his hand and release the arrow with directive purpose. The thinking, planning mind was regarded as an obstacle. American sportsman Catfish Hunter's description of his experience of 'the perfect game' further illustrates this point. He said: 'It was like a dream. I was going on like I was in a daze. I never thought about it the whole time. If I'd thought about it, I wouldn't have thrown the perfect game – I know I wouldn't.' Perhaps we can ask, if the thinking mind is no longer operational at these times, what is in control? How is the system being run? Whatever is in control not only appears to be much more efficient, but also the effect on consciousness seems to be extremely positive.

When we need fast reaction times, we need to leave our thinking mind behind. Fast and focused sporting responses can precipitate the shift. The result is fluid and accurate performance that is often accompanied by semi-mystical sensations of exhilaration leading to outright ecstasy, and even feelings of transpersonal unity. Basketball player Patsy Neal waxes lyrically about such experience:

'The athlete goes beyond herself; she transcends the natural. She touches a piece of heaven and becomes the recipient of power from an unknown source. The power goes beyond that which can be defined as physical or mental. The performance almost becomes a holy place – where a spiritual awakening seems to take place. The individual becomes swept up in the action around her – she almost floats *through the performance, drawing on forces she has never previously been aware of.'*

Perception may be altered too. A heightened sense of alertness is commonplace but there are reports of definite shifts in time and space. Some athletes claim to be able to perceive more detail and have a much more vivid field of vision than in ordinary states. This can be of great help. Objects can appear to be larger than they are in reality – to a golfer the hole can become as big as a wash tub so that it feels impossible to miss. Baseball players talk about how large the ball is when they are batting, and basketball players sometimes perceive the hoop as an ever expanding circle. That similar perceptual shifts are elicited by hypnosis indicates some factor in common. Does the root of these entire experiences lie in a shift towards right hemisphere dominance?

Seeing auras and energy are amongst the most extreme but most fascinating perceptual changes experienced in the field of sport. Such things are more often associated with yogis, saints and shamen, but American footballer, David Meggyesey accessed such perceptions after receiving a blow to the head during a practise game. As he sat on the sidelines in a semi-dazed state, he felt an eerie calm and beauty, and had impressions of outlines wavering gently in the fading light. This led to the perception of auras around some of the players. This opened the door to further experiences. In another game, he found himself playing in a kind of trance where he could sense the movements of the opposing players a split second before they happened. With this heightened sense of anticipation, he played a brilliant game. But was it anticipation?

Stunning research by Benjamin Libet of the University of California has indicated that brain activity begins some several hundred milliseconds *before* an individual is aware of deciding to act. This implies that our habitual consciousness lags behind the events of the world – a very problematical anomaly. What could be going on here? During our investigation into schizophrenia in the last chapter, we noted that voices heard in the head anticipated what the schizophrenic individual was attempting to put into words. Did something similar happen to David Meggyesey? These related phenomena can be best explained by the presence within all of us of a second consciousness system that operates in real time, (or even possibly outside linear time). It appears that this 'second self' can respond with greater speed and accuracy, which is why the aim of martial arts training is to reach this place beyond normal conscious action.

If our normal conscious self-system lags behind what is really going on, and makes up visual and mental compensatory links, the Hindu/Buddhist concept of Maya – the world of illusion – becomes more than a theoretical possibility. We really could be lost in the illusion of a supposed reality created by a damaged but dominant left-hemisphere self. A greater reality remains available to us but it is only accessed rarely. Murphy and White end their book with a parallel conclusion:

'...the very fact that such experiences arise spontaneously in many people suggests that they are a fundamental part of human nature. The fact that they burst in upon so many of us might indicate that we are designed to use them. Perhaps these strange abilities are part of a larger awareness and capacity that is pressing to be born. ... If a large-scale enterprise were mounted to explore these immense potentials of human nature, who knows what discoveries could be made. ... These extraordinary capabilities are probably only a glimmer of what human beings can achieve. We simply don't know the limits of long term research into these phenomena.'

We can only concur, with the proviso that we believe these abilities are not some new evolutionary advance waiting to happen. They are already part of us. They evolved with the rapid expansion of our human brain but somewhere along the road became progressively submerged. Today, accessing this second layer of function is a 'hit and miss' affair, but, with a sustained programme of research, a restoration of this lost area of consciousness may be possible. If we have suffered a degeneration of consciousness, what could be more important than ascertaining the extent of the problem and realising a solution?

SUMMARY ~ψ~

We have come to the surprising conclusion that our normal state of body/mind is not operating as well as it could and that a more functional state is hidden behind it. A critical look at areas such as human fertility, immunity and even baldness reveals evidence for a breakdown within these systems that appears to have been caused by an imbalance in steroids.

The most significant hormones in the story are testosterone and oestrogen. Either higher levels of these steroid hormones or, more accurately, an increase in their activity has precipitated a train of events that has upset optimal human performance. This has had profound repercussions on both our physical being and our consciousness, indeed down to the very way we think. The key steroid inhibitors – the heroes of our story – are melatonin, beta-carbolines and the vast numbers of chemicals, such as bioflavonoids, found in fruit. We are suffering today from a chronic shortage of these crucial chemicals, and, because of this lack, our pineal glands are not as active as they should be.

These ideas challenge many firmly held (but decidedly shaky) scientific theories that range from the descent of man to the rationality of our intellect; and even to what we should be eating. But together, all these anomalies inexorably point to the conclusions we have reached. There is firm evidence to support our thesis. We know from research that testosterone disproportionately affects the neural cells in the left hemisphere, and that melatonin suppresses the female fertility cycle. We also know that the thymus gland shrinks at puberty but, if steroids are suppressed, the thymus can reboot and greatly enhance the immune system. Put all the pieces of this jigsaw together and the picture we have been creating becomes clearly visible. Nobody has seen the relationship between all these factors before because they have been looking intently at the individual pieces of the jigsaw, not the whole picture.

A Native American Chief once told Carl Jung that he found white people to be always uneasy and restless. He could never fathom what these newcomers to the land of his forefathers wanted, or for what they were so restlessly seeking. He concluded that they were mad. He was not far off the mark. There is a profound discontentment in us that has resulted in an extraordinarily violent civilisation that has become a threat not only to itself but also to all life on this planet.

If we can recreate our ancestral hormonal environment through diet and a sustained reversal of cerebral dominance, a very different human may emerge. One with enhanced perception, a stronger immune system, more balanced dexterity, more efficient digestion and greater physical and mental capability. We would experience more profound and pleasurable sexuality too, coupled with a reproductive system that worked as nature intended. Even baldness would no longer be a problem. Most crucially, society would become much less aggressive and violent because our sense of self would change radically. Such a restoration of consciousness, with all these attendant benefits, is more that a theoretical possibility.

What is needed now is more research – diagnostic testing to confirm (or negate) our findings – and a multi-disciplinary investigation into ways to retrieve that which we have lost. We will investigate present cutting edge research and suggest new directions in our final chapter.

CHAPTER SIX

~ψ~

Mad Hatters and Maverick Scientists

We have considered and drawn on a wide range of detailed research work to support our new ideas of the human brain. While much of our synthesis appears speculative, our ideas are eminently testable and, as we shall see in this chapter, are supported by the work of leading researchers in their respective fields. To date we have had little access to suitable research laboratories but hope sometime in the future, as resources become available, to do some of this work. It would also be valuable to collaborate with others to discover whether these wonderful animal beings called *Homo sapiens* do indeed have greater potential than we presently realise. Here are a few avenues of research that we would like to follow up.

TRANSCRANIAL MAGNETIC STIMULATION ~ψ~

To directly test our hypothesis it would be necessary to reduce the influence of the left hemisphere for a period of time to see what would emerge. Fortunately, as we have seen earlier, it may be now possible to do this using Transcranial Magnetic Stimulation. It has been found that focusing a magnetic field on specific areas of the brain can interrupt that area's function. It would be most revealing to conduct trials using this technique in conjunction with the biochemical approaches we believe are needed to ensure that right hemisphere operation is not limited by deficiencies or impeded by a distorted balance of messenger hormones.

Although not following our criteria, initial work into this area has already been carried out in Australia by a team investigating aspects of autism. Robyn Young at Flinders University and Michael Ridding of the Royal Adelaide Hospital recently conducted an experiment to find out how applying a magnetic field to the left front temporal lobe would affect the ability of volunteers to carry out tasks. Following the ideas of Allan Snyder and

John Mitchell of the Centre for the Mind at the University of Canberra, they wished to test whether inhibiting this area of the brain would allow something like autistic savant skills to emerge.

Snyder's theory proposes that all of us have the ability to see and remember the world, like some savants, in incredible detail but 'high level processes' filter out the information before it is allowed to reach our conscious mind. Some now believe that autism is caused by a neurological condition in which a crucial part of the brain, which co-ordinates 'high level processes', does not develop correctly. Thus it is possible that these 'processes' keep us non-autistics from accessing what at times may be highly advantageous abilities. Snyder thinks that in savants the top level of mental processing – conceptual thinking and drawing conclusions – is somehow stripped away, and without it, savants can access a startling capacity for recalling endless detail or performing lightening quick calculations.

Although we find Snyder's ideas very relevant, according to our own theory it is the inadequacy of the dominant left hemisphere that, in normal individuals, is blocking these higher functions. In at least some autistic cases, developmental problems in the brain may have damaged the left hemisphere to such an extent that it is not able to completely dominate the right side and hence aspects of right hemisphere function can come through.

Darold Treffert, Clinical Professor at University of Wisconsin Medical School, has studied savant syndrome for over forty years. In a recent overview of the condition he considers, as others have proposed, that the extraordinary abilities may be due in part to 'paradoxical functional facilitation' of the right hemisphere, allowing for new skills as a compensatory process. But, as alluded to in the paper and confirmed in a recent personal communication, he is increasingly of the opinion that these right brain skills are not necessarily newly developed but instead represent latent dormant skills that emerge when released from, as he says, 'the tyranny of the left hemisphere'.

The results of the Young and Ridding experiment are consistent with this view. They indicate that using TMS to inhibit the left hemisphere can improve function. Some volunteers were reported to be better at calendar calculation whilst others improved accuracy in drawing a horse from memory. The subjects did not show anything as dramatic as the skills of autistic savants but two out of the seventeen volunteers displayed enhanced abilities and one improved in all the tests for memory, maths and art.

Whilst these results are a significant first step, in order to achieve the clear results we need to test our theory it would be necessary to change some of the parameters. We would not expect an astounding emergence of new abilities from experiments using normal subjects under normal conditions.

From our point of view it is incorrect to assume that humans more or less work as they should. This is an extremely important paradigm shift. Much research into our biology and psychology would benefit from building this wholly different assumption into the experimental method. If Ridding and Young's subjects had, prior to the TMS trials, engaged in practises that allowed more right hemisphere access, the repressed right hemisphere may have engaged more quickly and more fully when left hemisphere dominance was reduced. We would like to see comparative TMS trials conducted on a range of subjects with different diets (including those that approximate to our ancestral forest diet) and those who meditate regularly. We predict superior latent function would emerge in participants with better biochemistry and/or meditation practice. A greater number of skills should show up more quickly and more clearly in such subjects. The results of the Manchester sleep trial (see Chapter One) indicate that it was the underlying complex plant biochemistry in the participants' diet that allowed access to suppressed skills. The novel results came from this unique combination of diet and sleep deprivation.

Using TMS to investigate savant skills would be just a start. We could look for a whole host of enhanced functions such as increased dexterity and strength, as well as testing for increased brain coherence patterns. Although difficult to measure, the possible emergence of a different sense of self would be a fascinating element in this work. We could also test for such alternative abilities as remote viewing and telepathy (using two sets of TMS machines on two people). Putting in place the right biochemical parameters coupled with a reduced influence of the left hemisphere would be extremely revealing.

An important part of this research would be biochemical analysis. It is normally assumed that both hemispheres are adequate at what they do, and that they both display a degree of specialisation. The traits and abilities that have given rise to such notions are, we believe, better explained by a damaged left hemisphere with reduced abilities dominating a much more functional but inhibited right hemisphere. It follows from this, that there would be a different neuroendocrine system under right hemisphere control. If this were the case it would be eminently testable. For example, we would expect more melatonin circulating in the blood in a human system with reduced left hemisphere influence. Other biochemical assays to analyse changes in levels or activity of steroids would also be interesting.

A regime of biochemical testing was suggested but not carried out during the Manchester sleep deprivation experiment. If such assays had been made, we would have expected 'negative' changes initially because increasing tiredness in the left hemisphere, together with the physical trials, would have stimulated increased levels of stress hormones. Irritation often arises when we are tired – the left hemisphere self clearly becomes stressed and over-stretched when it is deprived of sleep. When this reaches a critical level, its function and dominance starts to fragment allowing the right hemisphere a freer rein. Thus, as the Manchester trial continued, a change to a generally lower level of circulating steroids with increased pineal activity would have been the likely response to the reducing influence of the left hemisphere. We predict there would be measurable endocrine changes (e.g. increased growth hormone and oxytocin) with a shift in dominance whether stimulated by TMS or sleep deprivation.

Speech has for a long time been recognised as the prime reason for believing the left hemisphere is specialised and highly functional. When the speech area of the left hemisphere is targeted by TMS, subjects find it difficult to speak. This would seem to confirm the left hemisphere's primacy in this area. Evidence too, from left hemisphere removal, indicates that there is very little right hemisphere speech in anyone beyond their childhood years, but perhaps the biochemical parameter mentioned above is a factor so far overlooked.

We have seen, in some circumstances, that the right hemisphere can handle speech and language very adequately. Even in later life, as A.W. seemingly experienced, an inhibited right hemisphere speech function can be activated with determination and focus. Using TMS to interrupt normal left hemisphere activity would allow this ability to be investigated thoroughly. We would not necessarily expect the function to activate immediately but, with repeated TMS sessions, some right hemisphere speech we predict would emerge.

If it were shown that the right hemisphere, free from the influence of the left, could adequately modulate speech then it would prove categorically that the left hemisphere is not overtly specialised. It would not only confirm that the right hemisphere is far more capable than generally accepted but also indicate that the left's 'specialist adaptations' were not much more than a response to a repeated pattern of use. It would also imply that the structural indications of enlarged areas such as Broca's are a response to continued usage rather than innate skill. Such an emergence of right hemisphere function would necessitate the rewriting of all the current textbooks on cerebral dominance.

TMS is potentially an extremely useful diagnostic tool with which to investigate the whole area of laterality and the nature of our different senses of self. Using the technique over extended periods could be interesting too. It is possible that such use would allow the right hemisphere to engage more fully.

It would also be fascinating to use TMS to investigate the phenomenon (see Chapter Five) of the time delay between the initial brain activity that indicates an action is about to happen and the later conscious decision to initiate the activity. It has been suggested that this bizarre scenario must have evolved because it benefits us in some way, but it seems more likely that it is yet another instance of dysfunction brought about by the rise of cerebral dominance. It is possible that there would be a more real-time consciousness under right hemisphere control: the delay could conceivably arise from the time it takes for the process to work from the right hemisphere self to the left, through the filter imposed by left hemisphere dominance. Perhaps martial artists and table tennis players, for example, short-circuit this process when they (semi-consciously) tap into more instantaneous function.

WEAK ELECTRO-MAGNETIC FIELDS ~ψ~

Another avenue of research, closely related to TMS work, concerns the effects of weak electro-magnetic fields. Back in the swinging sixties some intrepid mind experimenters induced hallucinations by wrapping a few strands of wire around their heads and running a small current through it. This they termed 'wire-heading'. Intrigued that some sort of field could apparently change perception, A.W. set up his own experiment (in 1997) using a wire coil and a battery set into a papier mache hat. Initially the effect was found to be weak but never-the-less perceptible. Various versions of the hat followed, some constructed with magnets and others with coils, and from trials with these it appeared that electro-magnetic fields rather than simple magnetic ones induced the perceptual effect.

Some initial research on superconductivity implies that an electro-magnetic field of up to eight gauss will enhance brain coherence but over anything over this has a negative effect. Using this information, a further hat was constructed that included a control for varying the field up to a maximum of eight gauss. A.W. reports that using the apparatus induced a sense of calm wellbeing that felt even better (euphoric) when combined with sleep deprivation and beta-carboline and/or neurotransmitter mimic supplementation. However in a 'normal' left hemisphere dominated state, the hat effect was barely noticeable, as it was to most people who tried the apparatus for the first time. It seems that an initial sensitivity was needed (access to the right hemisphere self) before the pleasant sensations were experienced. Continued practise helped increase this sensitivity but, as other trial participants have confirmed, combining the use of the hat with periods of sleep deprivation made the biggest shift. It was also found that, after being sensitised to the hat, it wasn't even necessary to put it on the head – just switching it on in the room would induce a powerful perceptual response.

Dr Michael Persinger, using equipment mounted in a modified motorcycle crash helmet has carried out more formal laboratory work on the perceptual effects of weak electro-magnetic fields. Persinger has been a professor at Laurentian University, Ontario, Canada since 1971 and in that time has written numerous papers and six books including 'The Neuropsychological Base of God Beliefs'. One of his most significant areas of research has been on the neural perceptual response to magnetic fields.

In 1987, he began systematically testing trial subjects with complex electro-magnetic fields after discovering that they could induce a number of experiences, ranging from a sensed presence to religious and mystical feelings. His aims were to determine which portions of the brain, or its electro-magnetic patterns, generate the experiences. The results of the experiments clearly show differences between the left and right hemispheres. Euphoric, religious or mystical experiences were accessed when the right hemisphere was stimulated. Furthermore, the sense of something or someone present seemed to be a left hemisphere interpretation of the intrusion from the right hemisphere. This 'sensed presence' was often interpreted (by the subject's left hemisphere) as spirits, ghosts, angels, gods, devils or aliens. Again it seems that the left hemisphere is trying its (limited) best here to attach some concept, however unlikely, to a numinous right hemisphere experience.

In Chapter Five, we discussed Dr Strassman's work on DMT. The perceptual responses elicited by this chemical are closely related to those found by Persinger. DMT may indeed be acting in a similar way to the electro-magnetic fields. Both may stimulate right hemisphere function leading to non-ordinary states of consciousness, which are then misinterpreted by the left. These 'thought' interpretations can be clearly seen to be false constructs for they depend on a cultural overlay. There were few, if any, cases of alien abductions in spacecraft before the popular explosion of sci-fi films and novels in the 1950s. Frightening perceptions that were once interpreted as visitations of the old hag, the incubus lover or demons are now more likely to be interpreted as visitations from the 'greys' or other species of outer planetary citizens. The underlying experiences may be identical but the explanations are coloured by the 'rational' mind. We get lost in the detail of the left hemisphere story rather than investigating the real perceptual experiences that are behind it.

Persinger has made some other fascinating connections between this type of response and electro-magnetism. He has convincingly shown that, in certain circumstances, the human perceptual equipment has an incredible sensitivity to electro-magnetic fields. Even a field generated by a digital clock/radio was found in one case to stimulate frightening nocturnal hallucinations. The girl in question was so terrified by her regular 'visitations' that she dreaded going to bed. Persinger investigated the case and found that the clock was producing a field. The clock was duly removed and her ghosts did not return.

When we are asleep the influence of our left hemisphere is reduced. This allows our right hemisphere, which requires less sleep, to function without so much interference. Our sensitivity to electro-magnetic fields is thus more apparent when we are asleep or in so called 'altered states' such as trance. This is significant. It is a further indication that the right hemisphere has greater functional sensitivity than the left.

Starting in 1992, individuals and even groups of people reported having religious experiences, specifically visions of Christ and Mary, at a site near Marmora in Ontario. These occurred near the top of a hill adjacent to a flooded magnetite mine. Between 1992 and 1997 epicentres for local seismic events periodically moved closer to the mine – and most of the experiences of spiritual beings occurred one or two days after such increases in geomagnetic activity. Persinger, fascinated by this correlation, concluded that the conditions produced by local geophysical and geological properties induced physiological changes within the thousands of people who visited the area. Direct measurements at the site indicated the presence of weak but complex magnetic fields that were similar to the experimental fields he had used to evoke altered states within the laboratory.

Such geomagnetic fields may provide an adequate explanation for the sensitive person's response to earth energy lines that link stone circles, ancient sites, churches and cathedrals. Even aliens and UFOs are 'seen' more commonly along geological fault lines and over areas of granite – places where there is greater electro-magnetic activity. Some people are naturally more sensitive to these electro-magnetic fields. For others it helps if the influence of our left hemisphere is reduced, but simply walking carefree in nature can help still the mind and allow more right hemisphere access. Some research even indicates that the focused electro-magnetic fields of such places as stone circles and cathedrals can stimulate the pineal. Thus it is probable that elevated levels of DMT may be involved in these mystical experiences.

The very versatile Persinger has also done some highly impressive work on memory. He, and his colleagues, have found that the storage of sensory information is lateralised to the right hemisphere. This fascinating finding raises questions about current notions of long and short-term memory. It is certainly not clear what the benefits or evolutionary pressures for such a system would be. Perhaps a more simple explanation is possible. Based on the evidence, presented throughout this book, for two selves emerging from structurally distinct hemispheres, we suggest there are two memories. Left hemisphere memory (roughly equivalent to the short term) is relatively poor in its depth, detail and recall. It is heavily biased towards the storage of labels, names, concepts and other verbal information. Right hemisphere memory, in contrast, is near complete. Events can be recalled in great depth and quality but access to this memory is usually blocked by the dominance of the left hemisphere self. What we usually remember is the left hemisphere's version of events (and as we have pointed out many times, the left hemisphere is very good at making up stories to fit its own comfortable agenda) but somewhere we do have more complete and realistic memory.

The fact that memory improving techniques, such as developing visual associations, relaxing and not actually focusing on the object of recall, all allow more right hemisphere access corroborates the right hemispheres primary role in memory. Furthermore, when our dominant self is out of the way, as it is during hypnosis and dreaming, we (that is our left hemisphere sense of self) has very little recollection of the activity. It is difficult for the left to remember anything when it is not fully present or dominant.

Finally, Persinger has seemingly shown that pleasure is mainly a right hemisphere activity. Low-energy magnetic fields applied over the right temporal lobe induced significantly more pleasurable feelings than when applied over the left. This makes sense to us. It is why we generally find it restoring to lose ourselves in such things as music and walking in beautiful surroundings. These pastimes calm the verbal side of our nature.

OTHER RESEARCH AREAS ~ψ~

Chimpanzees and Bonobos are our closest living relatives. In a comparative study of 97 key genes, it was found that 99.4 % of chimp and human genes are identical. The researchers suggested that the two could almost be reclassified as 'Homo'. This not only adds to the case that something other that genetic changes are responsible for the obvious differences between these apes and us, but also means that any biological correspondences could be more revealing than was previously thought. Investigations into such questions as the effect of diet on the fertility of bonobos and chimps and on their endocrine systems could tell us much about our own biology. Perhaps even behaviours, such as the aggressive monkey hunts carried out by young male chimps, could be partially explained by steroid activity. It is known that eating animal fat increases testosterone levels; thus including it in the diet could lead to more aggressive behaviour leading to more hunting activity. We could also ask whether giving extra melatonin to chimps would induce them to behave more like the bonobos. If so, it could indicate that our own behaviour and indeed our very sense of self is dependent on hormone balance which of course we believe relates to the balance between our right and left hemispheres. Norman Geschwind summarised this relationship.

'Unexpectedly, endocrinology and immunology appear to play a role in the determination of dominance, while conversely the dominance pattern may alter both endocrine and immune status.' (Cerebral Dominance, the Biological Foundations)

Simon Baron-Cohen's work, carried out at the Autism Research Centre, Cambridge, adds a significant twist to these ideas of hormonal balance. He analysed foetal testosterone in amniotic fluid and related the results to empathetic skills in the postnatal infants. He found that foetal testosterone shapes the neural mechanisms that underlie social development and has comprehensively shown that more testosterone acting on the foetus leads to less social skills. And there is a gender difference here too. Boys (male type brains) have less empathetic skills than girls do. Baron-Cohen has theorised that conditions like Asperger's syndrome and autism are extensions of the male type of brain caused by even greater testosterone levels acting on the foetus (extreme male brain theory). If this is so, then questions concerning the ordinary male type brain arise – could we be, in effect, damaged by the levels of testosterone we assume to be normal?

Testosterone itself does not do the work of masculising the brain: it is the steroid estradiol that plays the major role. But estradiol is made from testosterone by the activity of the enzyme aromatase. The variables that affect the degree of masculisation are, therefore, the amount of free testosterone available and the degree of aromatase activity. As we have already seen (see Chapter Three for the link with oestrogen dependent cancer), the activity of aromatase is inhibited by plant flavonoids and, more importantly, by melatonin. Less melatonin leads to more aromatase activity, which in turn leads to increased masculisation of the brain and, at the extreme end of the spectrum, autism (which appears to be becoming much more common). Our ancestral fruit-based diet would have been rich in aromatase inhibiting factors – and in the past our pineals would have pumped more melatonin too. The degree of masculisation of the male brain we see today, therefore, may well be an aberration that has had huge consequences for us.

We know that testosterone disproportionately retards the left hemisphere and we also know that elements occurring in our ancestral diets (flavonoids etc.) inhibit the activity of steroid hormones like testosterone. If we add these factors into the equation, it seems highly likely that particularly male brain development and hence social skills/behaviour are negatively affected by an over-exposure to testosterone, and that this is compounded by the dominance of the left hemisphere. Perhaps in a more natural state, our behaviour would have more in common with our gentle bonobo cousins than the aggressive chimps.

It would be very interesting to find out how different our human behaviour/social skills would be if the testosterone inhibiting influences, which were present in ancestral humans for millions of years, were reintroduced during uterine and post natal development. We know from Dr Richard Sharpe's work (see Chapter Three) that flavonoids alone can virtually stop the post-natal testosterone surge in males. If it were possible to correct the hormonal imbalance within pregnancy by incorporating such natural influences, the child would develop without the retarding influence of testosterone. This could be a crucial area of research.

Another related area of interest lies in the developmental process of the brain known as neural or synaptic pruning. It seems that as the brain grows, an excess of neural connections are made and these are selectively reduced from early post-natal development through to puberty. The exact mechanisms are not fully understood though the role of hormones such as testosterone and melatonin seem likely. If the hormone regime, that we believe is more natural, modified this pruning process, a different, perhaps more unified and cohesive, sense of self may result. This is possibly of great consequence.

Some unresolved questions relating to human fertility and development would be relatively straightforward to investigate. For example, do mothers who meditate have more melatonin in their breast milk? As meditation engages more right hemisphere function, and stimulates greater pineal activity, we would expect this to be so. Hajime Kimata, of the Department of Allergy, Moriguchi-Keijinkai Hospital, Osaka, has already found that laughter increases the levels of melatonin in breast milk. Nursing mothers who were shown a humorous DVD produced more melatonin than those who were shown dull weather information. Laughter (like meditation) is indicative of a shift in cerebral dominance from left to right, and when the right hemisphere is engaged in this way it stimulates greater pineal activity.

Evidence from animal research indicates that there is a link between the duration of breast-feeding and the timing of puberty. We expect tests on humans would show that a longer period of breast-feeding results in a later puberty, and that this relates to relatively higher levels of melatonin expression/pineal activity. As we discussed in earlier chapters, a longer juvenile period could be beneficial to neural development and conversely a shorter period detrimental. This research therefore is not merely academic; the shortening of the human juvenile period that is continuing apace today could be having serious consequences.

Although there is still a need for the final diagnostic research to be completed, it appears that meditation/laughter/right hemisphere activity leads to more melatonin production and to feelings of greater well-being. Work already carried out in 2007 at the Pasteur Institute in Paris (Thomas Bourgeron et al) has found that individuals with autism spectrum disorders have lower than normal levels of melatonin. The French team highlight the crucial role of melatonin in human cognition and behaviour. They hypothesize that low melatonin has a direct effect on the modulation of neural networks and can amplify the effect of other genetic mutations in autistic individuals.

It is now clear that less melatonin is associated with more testosterone (steroid hormones), greater masculisation of the brain (in both sexes), which in extreme cases leads to autism, and a general tendency towards aggressive and paranoid behaviour. Tony Wilson, of Wake Forest University (2007), has discovered too that a correlation exists between autism and deficiencies in the left hemisphere. All this, coupled with the finding that microtubules (see Chapter Five) are extremely responsive to melatonin levels, indicates how crucial a chemical melatonin is to us. As melatonin production declines with age (as left hemisphere dominance increases), the structure of our microtubules and any role they play in cognition or consciousness will also change. If we are all suffering from a chronic deficiency of melatonin and this worsens with age the consequences on our perception and behaviour may be far more severe than is currently recognised.

Melatonin's modulating effect on steroids makes it a key variable in neural development and all steroid-related function. However, we now know that it can also act as a transcription factor in its own right – it is involved directly in the regulation of protein structure as well as indirectly via its modulation of steroids. This may turn out to be another reason why melatonin is so important, and indeed why a low expression of melatonin is associated with such illnesses as autism and Alzheimer's.

There have been tens of thousands of studies on human health. Billions of pounds are spent on drug and food research, and on health services to bail out ill people. Despite this, conclusive research on which foods are really the best for us has not been done. Study after study has proved the link between some foods and much ill health, including cancer, but no governmental body will categorically say that many elements of our modern day diet are damaging (though they are beginning to get close). It would be a radical but realistic approach to conduct future studies on food with the underlying assumption that we are basically primates. Our closest relatives eat a very high percentage of fruit and vegetable matter and eat it uncooked. Investigations into the most beneficial diet for humans should at least take this point into consideration. Dr Katherine Milton's paper (2003) on the micronutrients in a primate diet has highlighted the abundance of high quality nutrients provided by a forest diet. This contrasts sharply with the relatively poor nutrient levels in our own diets. However nutritionalists tell us that vegetables, fruits, nuts and seeds that are the mainstays of raw and vegetarian preferences have enormously higher nutrient values than foods such as potatoes, wheat and rice. These latter foods, that are usually cooked, provide the bulk of most modern diets and are directly related to our nutritional impoverishment.

Despite the routine over-processing, our food industries are beginning to wake up to the benefits of plant chemicals, and particularly their ability to regulate cholesterol. They have found that including them in foods can be a marketing trump card. Thus we now see, for example, phytosterols added to margarine and yoghurt 'to fight cholesterol' – a far cry from ingesting such factors via wild figs and mangoes but at least a minor step in the right direction. There is also an increasing appreciation of the importance of fatty acids. Studies have shown that children with developmental coordination disorder and attention deficit problems dramatically improve when treated with essential fatty acids. Children taking EFAs in pill form have also shown outstanding leaps in mental development. In a detailed study of four children between the ages of eight and 13, brain scans revealed three year's worth of development in just three months and, during the trial, they displayed remarkable improvements in reading, concentration, problem solving and memory – even their hand writing became neater and more accurate. One boy, who previously avoided books and was hooked on TV, became engrossed by written stories and declared he was 'bored' of television. Professor Puri and his team at St George's Hospital have concluded that this research shows just how much harm junk food is doing to our children's development.

Fatty acids themselves are highly volatile and will oxidise even at room temperature. Such oxidisation will substantially destroy both their integrity and functionality. Heating them to high temperatures (in cooking) will destroy them even more comprehensively. As these EFAs are the building blocks of our brains and nervous systems, their importance cannot be over-emphasised. In biological systems they are heavily protected with anti-oxidants. It is significant that the children in the trial, who showed so much functional improvement, took their fatty acids in 'natural' capsule form. Nutritional advice derived from such studies should be widely disseminated. Although more research is always needed, we now know enough for the general public to be given clear guidance to the foods that are really needed to enhance our mental and physical development.

There are many other avenues of research that we would like to follow up. For instance, it would be fascinating to investigate the differential in requirement for sleep between the two hemispheres and also delve into any differential in perceptual response too. If, as we predict, major differences were demonstrated it would be clear that there was more to the laterality enigma than is generally recognised.

There are many anomalies which are evident in the human make up – cerebral dominance, handedness, questions around sleep and hypnotism – but orthodox science hasn't addressed the questions that these anomalies pose in anything like a comprehensive or integrated way. Scientists have looked at isolated details but not stepped back to see the bigger picture. Our understanding and depth of knowledge of ourselves has increased exponentially over the last hundred years but this knowledge is squeezed into the accepted paradigm that humans are at the pinnacle of evolution and development. By questioning this paradigm and asking instead whether there has been a stalling or glitch in our development, a fresh light can be shone on many of these so called anomalies.

In contrast, spiritual traditions, though using a very different language, have always looked at the 'fallen' side of man. They have devised widespread practices for quietening the left hemisphere's sense of self to allow more right hemisphere perception to emerge. This is equated with a sense of connection and oneness that is at the heart of religious experience. Modern science has looked very successfully at the mechanics of the instrument that generates the sense of self, everything from the neurones to neurotransmitters, but religions have always been concerned with the 'music that the instrument plays'. It could be time to combine these approaches. We are now in a unique position to investigate such fundamental questions as 'who are we'? If our usual sense of self is actually, as the ancient traditions have suggested for thousands of years, a consequence of a flaw in our perception, then there is a real possibility of making a highly significant breakthrough. The evidence and framework we have presented here could represent a step in this direction.

RESTORATION ~ψ~

Pointers from the research work of leading scientists like Simon Baron-Cohen, Allan Snyder, Vilayanur Ramachandran, Michael Persinger, Katherine Milton and many others have led us to the conclusion that we are suffering from a neurological condition arising from an abnormal level of testosterone activity, impinging most crucially during our natal and post-natal development. As the symptoms of this condition manifest in our sense of self, how we feel and how we relate to others, the consequences are far-reaching. Baron-Cohen has shown that the male pattern brain, which is the result of this condition, lacks empathy and social skills. And it is this type of brain that is responsible for all the social unrest, wars and brutality that has been rampant particularly since patriarchal culture took hold (itself a consequence of increasing left hemisphere dominance).

Such behaviours are not ultimately the result of greed, desire or an evil nature but of the way our perceptual systems work. Under a different hormonal regime our brains would develop in a different way resulting in a different outlook and a different sense of self. We would naturally relate in a harmonious way. With an increased background sense of wellbeing and free from our underlying fears and insecurities, we would awaken from the nightmare of violence that has been created from left hemisphere confabulation. Such violence to our fellow man, and even the damage inflicted upon our planet, would be impossible if our brains were soft-wired for empathy and not hard-wired for war.

The problem

In a nutshell – one half of our consciousness facilitating system is structurally damaged. This severely limits its perception and paradoxically, due to the detrimental psychological changes resulting from the damage, it always takes control. The other half is suppressed, and both halves have been built from less than optimal materials. The left hemisphere – the dominant side of the brain that modulates our sense of self and how we act –

185

no longer has the capacity to experience reality directly. Instead it builds its identity from ideas, concepts and early experiences, and, once this has been established, it has great difficulty updating its interpretation of reality, even in the face of directly conflicting evidence. Compared to the right hemisphere, the left has limited perceptual abilities and the testosterone-inflicted damage has further reduced its capacity in areas such as memory, joy and empathetic connection.

The right hemisphere remains relatively undamaged. It still experiences 'real time' reality and can respond fluidly and appropriately to ever changing circumstances. Key evidence also indicates that the right hemisphere possesses enhanced functions. These are usually regarded as extraordinary as they only emerge in exceptional circumstances (savant syndrome, child prodigies, brain damage). That these circumstances all involve a reduction in the influence of the left hemisphere suggests that right hemisphere function is latent and held in check.

The primary limiting factor is cerebral dominance. The inhibiting effects of the left hemisphere, which routinely suppress the right's mode of operation, are extremely difficult to circumvent. The second limiting factor is a chronic deficiency of the complex and specific neurochemistry essential for developmental construction and optimal functioning. These limiting factors would need to be addressed simultaneously to enable something approaching a restoration of our consciousness system.

A Potential Solution

Although our conclusions appear dramatic and audacious, identifying a problem is the initial and very necessary stage in finding a solution. As the solution in this case entails nothing less than a restoration of consciousness, this could be regarded as a very exciting prospect. But is this possible, and what would we be like if the hormonally related glitch could be rectified?

Not only do we believe a restoration is possible but also, if we look behind the outer aspects of religion and myth, we find that a search for a solution has been, either intuitively or more directly, one of the preoccupations of man for millennia. We feel humankind would be happier and much less fearful, devious and violent if we could regain our lost perceptual heritage. Additional benefits would include a more vigorous immune system, greater physical strength and enhanced perceptual abilities.

To achieve such desired results, we would need to minimise the negative effects of the left hemisphere and then retrieve as much latent ability as we can from the right. Lifting dominance can, under certain circumstances, release further function from the right hemisphere. However, re-establishing the fullest potential would require the biochemistry needed for optimal functioning and the raw materials that would allow the system to rebuild. This would require a primate-like diet, rich in fatty acids and the potent chemicals found within fruit.

Raising the quality of neural construction materials should encourage the internal production of chemicals that would further contribute to the restoration process. For example, the pineal would partially re-engage and begin to produce more melatonin. These materials (MAO inhibitors and flavonoids) directly elevate pineal activity, and, as we have already noted, this gland becomes more active when left hemisphere suppression is reduced.

Reducing sleep, meditation and biochemical supplementation would all be easily accessible parts of a package of approaches to help restoration. But the one difficulty that remains is how to keep the dominant left hemisphere muted for long enough to allow our second system to become fully functional. Many people meditate and some get glimpses of higher function. Some people eat a diet that may be suitable for restoration, and they too may get glimpses or become more 'sensitive'. Occasionally an individual may do both, but a sustained effort is needed to make big shifts using these methods. Not many people are able or prepared to do this, particularly in our modern, rushed and demanding world.

Other 'high tech' approaches could perhaps offer some help. Transcranial Magnetic Stimulation offers us the opportunity to suppress the activity of the left hemisphere and allow the right to emerge from its shadow. Using a machine that could minimise the disruptive effect of the left hemisphere for a period of time could be very helpful and extremely interesting.

To summarise then, the restoration of an individual's consciousness system would need to include a combination of the following approaches:

1. Rebuilding and restoring the neural structure by including the important nutritional elements of a primate-like diet.
2. Putting in the optimal biochemistry to lift function: this would include monoamine oxidase inhibitors, such as those found in passion flower tea and figs, and also melatonin to boost the pineal and inhibit steroids. There may even be a need for chemicals such as DMT (at a clinically determined level) to re-activate the second system.
3. Engaging in techniques such as meditation and sleeping less to reduce left hemisphere dominance.
4. Possible short-term use of high tech processes like Transcranial Magnetic Stimulation to allow the right hemisphere some time free from suppression.

This combination of approaches would stimulate the pineal to produce more melatonin, pinoline and possibly DMT. More melatonin means greater suppression of steroids, like testosterone, and this will have two further effects. It will block or reduce the ongoing damage caused by too much steroid activity and will lift the dampening effect that steroids have on neurotransmitter activity. This would initiate a reversal of the neuro-endocrine damage that began in those distant days when we were cut off from the forest biochemistry.

At times, even with the consciousness system that we have now, we can attain inspirational and blissful states of mind. We can only wonder what we would be like with a consciousness system that was fully engaged. If research were to solve this most central of human problems, we would expect profound changes. Our damaged consciousness systems could not be restored without altering our present consciousness. We would be different. Hopefully these changes would be hugely beneficial not only to us as individuals but also to our whole global community.

If the model outlined really is correct then, like the anosognosia patients told of their paralysis, just being aware of the facts is rarely enough. It takes something else – an experiential shift, a eureka moment of profound realisation – basically an intervention of the right hemisphere that turns the facts into experiential reality. It is not the left hemisphere self that is in the dark, it is the self we refer to as I, me and you.

This book has been written by the left brain for the left brain. For those amongst (or within) us who prefer an abridged, more pictorial right brained version – see overleaf.

Any Change in Neural Structure or Neurochemistry Alters Consciousness

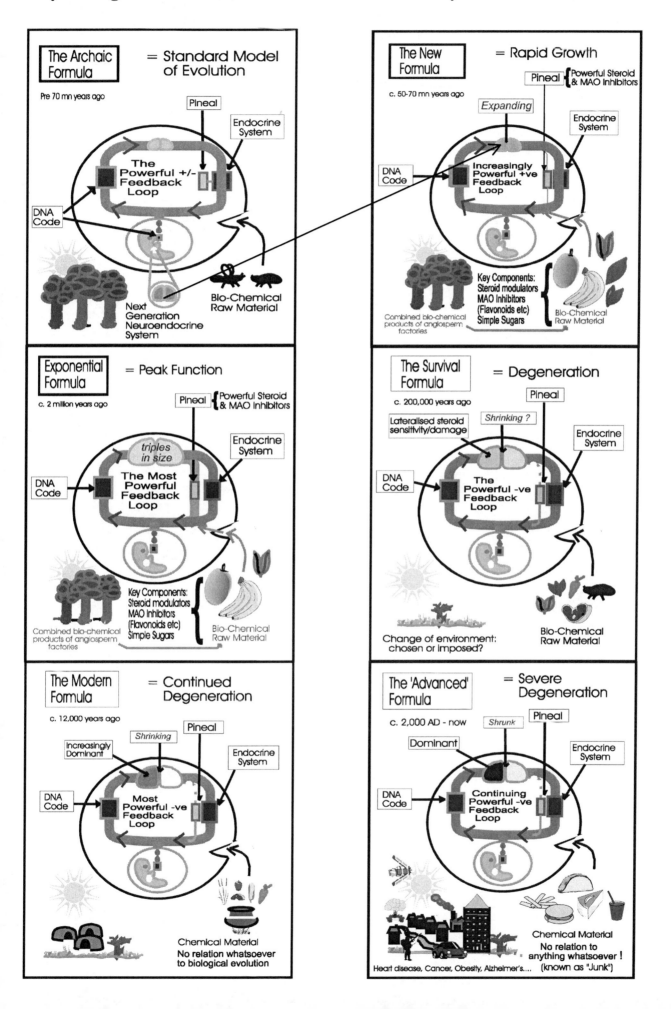

Recommended Reading

~ψ~

Charter, Steve. *Sustainability, Consciousness and Climate Change,* Lulu Books 2007

Geschwind, Norman and Galaburda, Albert. M. *Cerebral Dominance,* Harvard University Press, 1984

Heinberg, Richard. *Memories and Visions of Paradise,* Quest Books 1995

Jaynes, Julian. *The Origin of Consciousness in the Breakdown of the Bicameral Mind,* The Penguin Press 1979

Lancaster, Brian. *Mind, Brain and Human Potential*, Element 1991

Morgan, Elaine. *Scars of Evolution*, Penguin Books, 1991

Murphy, Michael and White, Rhea. A. *In The Zone.* Arkana 1995

McTaggart, Lynne. *The Field*, HarperCollins 2001

Ornstein Robert. *The Right Mind,* A Harvest Book Harcourt Brace & Company 1997

Pedersen, David. L. *Cameral Analysis*, Routledge 1994

Pierpaoli, William, Regelson, William and Coleman, Carol. *The Melatonin Miracle,* New York: Simon and Schuster, 1995

Price, Weston, A. *Nutrition and Physical Degeneration*, Price-Pottenger Nutritional Foundation, La Mesa, California, 1970

Strassman, Rick. *DMT the Spirit Molecule*, Park Street Press, 2001

References

~ψ~

Introduction

Allan, D.S. and Delair, J.B. *When the Earth Nearly Died*, Gateway Books, 1995
De Santillana, Giorgio and Von Dechend, Hertha. *Hamlet's Mill*, Gambit, Boston, 1969
Dupuis, Charles. *Origine de tous les Cultes et toutes les Religions*, Paris , 1795
Flem-ath, Rand and Rose, *When the Sky Fell*, Weidenfeld and Nicolson, 1995
Gimbutus, Marija. *The Living Goddesses*. University of California Press, 1999
Hapgood, Charles. *Maps of the Ancient Sea Kings*, 1966, reprint Adventures Unlimited Press, 1996
Hancock, Graham. *Fingerprints of the Gods*, Heinemann, 1995
Hancock, Graham. *Underworld- Flooded Kingdoms of the Ice Age*, Penguin/Michael Joseph, 2002
Heinberg Richard. *Memories and Visions of Paradise* Quest Books 1995
Knight, Christopher and Lomas, Robert, *Uriel's Machine*, Century Books, 1999
Oppenheimer, Stephen. *Eden in the East – The Drowned Continent of Southeast Asia*, Weidenfeld and Nicolson, 1998

CHAPTER ONE: **Two Sides to Everything**

Blackmore, Susan. 'Why Psi Tells Us Nothing About Consciousness' in *Toward a Science of Consciousness II. The Second Tuscon Discussions and Debates*, The MIT Press 1998
Bolduc Christianne, Daoust Anne-Marie, Limoges Elyse, Braun Claude M.J, Godbouta Roger. 'Hemispheric lateralization of the EEG during wakefulness and REM sleep in young healthy adults,' *Brain and Cognition 53* 2003
Carter, Rita. 'Tune In????' *New Scientist,* 9th October 1999
Drummond Sean P. A, Brown Gregory G, Gillin J, Stricker John L, Wong Eric C, Buxton Richard B. 'Altered brain response to verbal learning following sleep deprivation', *Nature Vol 403*, February 2000
Edwards, Betty. *Drawing on the Right Side of the Brain*, Los Angeles: J.P. Tarcher Inc, 1979
Elkin, A.P. *Aboriginal Men of High Degree*, St Lucia, Queensland: University of Queensland Press, 1980
Geschwind, Norman and Galaburda, Albert. M. *Cerebral Dominance* Harvard University Press, 1984
Geschwind, N. and Behan, P.O., 'Laterality, Hormones and Immunity' in *Cerebral Dominance,* Harvard University Press, 1984
Govinda, Lama Anagarika. *Foundations of Tibetan Mysticism*, Rider 1969
Grinspoon, L., and Bakaler, J.B. 'Marijuana as Medicine', *Journal of the American Medical Association*, 1995
Harney, Bill. *To Ayers Rock and Beyond,* Ian Drakeford Publishing, 1988

Jackson, Deborah. 'Heard, and Seen', *Natural Parent*, July/August 2001

Jaynes, Julian. *The Origin of Consciousness in the Breakdown of the Bicameral Mind*, The Penguin Press 1979

Joudry, Patricia. *Sound Therapy for the Walk Man*, Steele and Steele, 1984

Lancaster, Brian. *Mind, Brain and Human Potential*, Element 1991

Landolt Hans-Peter, Ph.D., and Gillin J. Christian, M.D. 'Different Effects of Phenelzine Treatment on EEG Topography in Waking and Sleep in Depressed Patients', *Neuropsychopharmacology 27:462–469*, 2002

Lawlor, Robert. *Voices of the First Day – Awakening in the Aboriginal Dreamtime,* Inner Traditions, Vermont, 1991

Lissoni, P., Resentini, M., and Fraschini, F. 'Effects of Tetrahydrocannabinol on Melatonin Secretion in Man', *Hormone and Metabolic Research*, 1986

Morris, K. 'The Cannabis Remedy – Wonder Worker or Evil Weed', The Lancet, Vol 350, December 20, 1997

Motluk, Alison. 'When half a brain is better than one' *New Scientist Vol 150 issue 2026 April 1996*

Pedersen, David. L. *Cameral Analysis*, Routledge 1994

Propper RE, Lawton N, Przyborski M, Christman SD.'An assessment of sleep architecture as a function of degree of handedness in college women using a home sleep monitor*', Brain Cogn.* Apr 2004

Raikov, VL. 'EEG recordings of experiments in hypnotic age regression', *Imagination, Cognition and Personality*, 3, 1983

Reiter, R.J. and Robinson, J. *Melatonin: Your Body's Natural Wonder Drug*, Bantam Books 1995

Saddhatissa, H. *The Life of the Buddha*, Allen and Unwin 1976

Serafetinides, E.A. 'The Significance of the Temporal Lobes and of Hemispheric Dominance in the production of the LSD-25 Symptomatology in Man', *Neuropsychologia, Vol 3*, 1965

Sheldrake, Rupert. *The Sense of Being Stared At,* Crown Publishing, 2003

Shlain, Leonard. *The Alphabet Verses The Goddess,* Penguin 1999

Sperry, R.W. 'Hemispheric deconnection and unity in conscious awareness', *American Psychologist 23*, 1968

Sperry, R.W. *'Lateral Specialisation of Cerebral Function in the Surgically Separated Hemispheres'*, The Psychophysiology of Thinking, 1973

Wilson, Robert, Anton. *Sex and Drugs: A Journey Beyond Limits*, New Falcon Publications 1973

Wright, Tony. The expansion and degeneration of human consciousness, *Informatica Volume 22 Number 3, 1998*

CHAPTER TWO: **From the Forest**

Andlauer W, Stumpf C, Hubert M, Rings A, Fürst P. 'Influence of Cooking Process on Phenolic Marker Compounds of Vegetables', *Int J Vitam Nutr Res,* Mar 2003

Batmanghelidj, F. *Your Body's Many Cries for Water*, Tagman Press, 2000

Best, Simon, 'A Nutritional Approach to Treating ADHD', *Nexus*, Vol. 8 No. 6, Oct 2001

Brookes, Martin. 'Apocalypse then', *New Scientist*, 14th August 1999

Blaut M., Schoefer L., Braune. A. 'Transformation of Flavonoids by Intestinal Microorganisms', *International Journal for Vitamin and Nutrition Research, Seite 79 - 87, Band 73,* 2003

Colgan, M. *Your Personal Vitamin Profile*, Quill: New York, 1982

Courtillot, Vincent. *Evolutionary Catastrophes*, Cambridge University Press, 1999

Darwin, Charles. *The Descent of Man*, 1871

Davies, David. *Centenarians of the Andes,* Anchor Press, 1975

Fontana, L. Shew, J.L. Holloszy, J.O. and Villareal, D.T. 'Low Bone Mass in Subjects on a Long-term Raw Vegetarian Diet', *Arch Intern Med*, Vol.165, Mar 28, 2005

Fox, Douglas. 'Cut the Carbs', *New Scientist*, 18th March 2000

Herraiz, Tomas. 'Analysis of the bioactive alkaloids tetrahydro-B-carboline and B-carboline in food', *Journal of Chromatography A, 881,* 2000

Kaplan, Matt. 'Make love, not war' (bonobos), *New Scientist,* Dec 2006

Kapleau, Phillip. *To Cherish All Life*, The Zen Center, 1981

Kenton, Leslie and Susannah, *Raw Energy*, Century 1984

Khamsi Roxanne.'You are what your grandmother ate', *NewScientist.com news service 22:00 13 November 2006*

Koestler, Arthur. *The Ghost in the Machine*, Macmillan, 1968

Kouchakoff, Paul. 'The Influence of Food Cooking on the Blood Formula of Man', *Proceedings of First International congress of Microbiology,* Paris, 1930

Kuratsune, Manasore. 'Experiments of Low Nutrition with Raw Vegetables', in *Kyushu Memoirs of Medical Science*, Vol. 2, No 1-2, June 1951

Leakey, Richard. *The Making of Mankind*, Michael Joseph Limited, 1981

Lewin, R., 'Rise and Fall of Big People', *New Scientist*, Vol. 146, 22nd April 1995

Mayell Hilary. 'Oldest Human Fossils Identified', *National Geographic News,* February 2005

Milton, Katherine. 'Diet and Primate Evolution', *Scientific American*, August 1993

Milton Katharine PHD. 'Nutritional Characteristics of Wild Primate Foods: Do the Diets of Our Closest Living Relatives Have Lessons for Us?', *Nutrition Vol. 15, No. 6,* 1999

Milton Katharine. 'Micronutrient intake of wild primates: are humans different?' *Comparative Biochemistry and Physiology Part A 136 47-59,* 2003

Morgan, Elaine. *Scars of Evolution*, Penguin Books, 1991

Odent, Michael. *Primal Health,* Century Hutchingson Ltd, 1986

Phillips, Roger and Rix, Martyn. *Vegetables*, Pan Books 1993

Pottenger, F.M.Jr. *Pottengers's Cats*, Price-Pottenger Nutritional Foundation, La Mesa, California, 1983

Pottenger, F.M.Jr. and Simonsen, D.G. 'Heat Labile Factors Necessary for the Proper Growth and Development of Cats', *Journal of Laboratory and Clinical Medicine,* Vol. 25, 1939

Powell, C.S. and Gibbs, W.W. 'Rambling Road to Humanity', *Scientific American*, 16th June 1997

Price, Weston, A. *Nutrition and Physical Degeneration*, Price-Pottenger Nutritional Foundation, La Mesa, California, 1970

Raichle Marcus E., and Gusnard Debra A. 'Appraising the brain's energy budget', *PNAS August 6, vol. 99 no. 16 10237-10239,* 2002

Tattersall, Ian. 'Out of Africa Again ... and Again?, *Scientific American*, April 1997

Tattersall, Ian. 'Once We Were Not Alone', *Scientific American,* January 2000

Williams, R. *Nutrition against Disease*, Pitman Publishing Co, New York, 1971

CHAPTER THREE: **Figs, Steroids and Feedback Loops**

Balch, P.A and J.F. *Prescription for Nutritional Healing,* Avery/Penguin Putnam 2000

Bentley, George. E., Van't Hof, Thomas, J. and Ball, Gregory, F. 'Seasonal neuroplasticity in the songbird telencephalon: A role for melatonin', *John Hopkins University* 1999

Brookes, Martin. 'Apocalypse then', *New Scientist,* 14th August 1999

Campbell, Don. *The Mozart Effect,* Hodder and Stoughton 2001

Campbell DR, Kurzer MS.'Flavonoid inhibition of aromatase enzyme activity in human preadipocytes.' *J Steroid Biochem Mol Biol,* Sept 1993

Cos Samuel, Martínez-Campa Carlos, Mediavilla Maria D. and Sánchez-Barceló Emilio J. 'Melatonin modulates aromatase activity in MCF-7 human breast cancer cells', *Journal of Pineal Research Volume 38,* March 2005

D'Adamo Peter, N.D. 'Aromatase Inhibitors And Estrogen Dominance', Epub – dadamo.com, 2004

Grosvenor, Clark. 'Hormones and Growth Factors in Milk', *Endocrine Reviews,* Vol.14, No.6, 1992

Fox, Douglas. 'Cut the Carbs', *New Scientist,* 18th March 2000

Haldar C, Fukada Y, Araki M. 'Effects of gonadal steroids on pineal morphogenesis and cell differentiation of the embryonic quail studied under cell culture conditions', *Brain Res Dev Brain Res,* Oct 2003

Jaynes, Julian. *The Origin of Consciousness in the Breakdown of the Bicameral Mind,* The Penguin Press 1979

King TS, Steger RW, Richardson BA, Reiter RJ. 'Interaction between pargyline, a monoamine oxidase inhibitor, and beta-adrenergic receptors in the rat pineal gland', *Prog Clin Biol Res.* 1982

Lampe Johanna W. Gustafson Deborah R. Hutchins Andrea M. Martini Margaret C. Li Sue. Wahala Kristiina. Grandits Greg A. Potter John D. and Slavin Joanne L. 'Urinary Isoflavonoid and Lignan Excretion on a Western Diet: Relation to Soy, Vegetable, and Fruit Intake', *Cancer Epidemiology, Biomarkers & Prevention Vol. 8, 699–707,* August 1999

LeMay, Marjorie. 'Radiological, Developmental and Fossil Asymmetries', *Cerebral Dominance,* Harvard University Press, 1984

Lewin, R., 'Rise and Fall of Big People', *New Scientist,* Vol. 146, 22nd April 1995

Martini, Frederic. *Fundamentals of Anatomy and Physiology,* Simon and Schuster 1998

Mckenna, Terence. *True Hallucinations,* HarperCollins (Rider), 1993

McTaggart, Lynne. *The Field,* HarperCollins 2001

Middleton Elliott, Jr. Kandaswami Chithan, And Theoharides Theoharis C. 'The Effects of Plant Flavonoids on Mammalian Cells: Implications for Inflammation, Heart Disease, and Cancer', *Pharmacological Reviews Vol. 52, No. 4,* 2000

Miller Helen L, Ekstrom R David, Mason George, Lydiard R Bruce, Golden Robert. 'Noradrenergic Function and Clinical Outcome in Antidepressant Pharmacotherapy', *Neuropsychopharmacology,* 2001

Milton, Katherine. 'Diet and Primate Evolution', *Scientific American,* August 1993

Nair N.P.V. M.D., Ahmed S.K., M.D., Ng Ying Kin N.M.K., Ph.D. 'Biochemistry and Pharmacology of Reversible Inhibitors of MAO-A Agents: Focus on Moclobemide', *J Psychiatr Neurosci, Vol. 18, No. 5,* 1993

Pert, Candace. B. *Molecules of Emotion,* Simon and Schuster, 1998

Pierpaoli, William, Regelson, William and Coleman, Carol. *The Melatonin Miracle*, New York: Simon and Schuster, 1995

Price, Weston, A. *Nutrition and Physical Degeneration*, Price-Pottenger Nutritional Foundation, La Mesa, California, 1970

Roney-Dougal, S.M. 'Recent Findings Relating to the possible Role of the Pineal Gland in Affecting Psychic Ability', *Journal of the Society for Psychical Research*, Vol. 55, No. 815, 1989

Roney-Dougal, S.M. and Vogl, G. 'Some Speculations on the Effect of Geomagnetism on the Pineal Gland', *Journal of the Society for Psychical Research*, Vol. 59, No. 830, 1993

Rosenberg Zand Rachel S, et al. 'Flavonoids can block PSA production by breast and prostate cancer cell lines', *Clinica Chimica Acta Volume 317, Issues 1-2,* March 2002

Rosenberg Zand Rachel S, et al. 'Flavonoids and steroid hormone-dependent cancers', *Journal of Chromatography B Volume 777, Issues 1-2,* 25 September 2002

Salib, Michael. Sexual Dimorphism in the Human Brain, 1997

Sanderson JT, Hordijk J, Denison MS, Springsteel MF, Nantz MH, van den Berg M. 'Induction and inhibition of aromatase (CYP19) activity by natural and synthetic flavonoid compounds in H295R human adrenocortical carcinoma cells', *Toxicol Sci,* 2004

Sharpe Richard M, Martin Bronwen, Morris Keith, Greig Irene, McKinnell Chris, McNeilly Alan S. and Walker Marion. 'Infant feeding with soy formula milk: effects on the testis and on blood testosterone levels in marmoset monkeys during the period of neonatal testicular activity', *Human Reproduction, Vol. 17, No. 7, 1692-1703,* July 2002

Tattersall, Ian. 'Once We Were Not Alone', *Scientific American,* January 2000

Taylor Loverine, and Grotewold Erich. 'Flavonoids as developmental regulators', *Current Opinion in Plant Biology Volume 8, Issue 3,* June 2005

Wrangham, Richard and Peterson, Dale. *Demonic Males – Apes and the Origins of Human Violence,* Houghton Mifflin, 1996

Wright, Tony. *'Tropical forest biochemistry, the driving force in human evolution'*, Epub – grahamhancock.com 2006

Zimmerman, Marcia C.N. *'Phytochemicals: Nutrients Of The Future',* Epub – realtime.net

CHAPTER FOUR: **Consciousness**

Atwater, P.M.H. *Beyond The Light,* Birch Lane Press 1994

Bandler, Richard and Grindler, John. *Frogs into Princes,* Real People Press, 1979

Boatman Dana, Freeman John, Vining Eileen, Pulsifer Margaret, Miglioretti Diana, Minahan Robert Carson Benjamin, Brandt Jason, and McKhann Guy. 'Language Recovery after Left Hemispherectomy in Children with Late-Onset Seizures', *Annals of Neurology Vol 46 No 4* October 1999

Baker, John R. Consciousness Alteration as a Problem-Solving Device: The Psychedelic Pathway, *Yearbook for Ethnomedicine and the Study of Consciousness Issue 3, 1994*

Bruner Jerome S. and Postman Leo. 'On the Perception of Incongruity: A Paradigm' Harvard University First published in *Journal of Personality, 18, 206-223.* (1949)

Caffell, Colin. 'A Dragon Called Fear', *Caduceus,* 33, Autumn 1996

Casteneda, Carlos. *Journey to Ixtlan,* Penguin Books 1974

Clamp, John. 'High on Life – an interview with Richard Bandler', *Kindred Spirit* 37, Winter 1996/7

Claxton, Guy. *Hare Brain, Tortoise Mind: Why Intelligence Increases When You Think Less*, Forth Estate, 1997

Curtiss Susan, De Bode Stella. 'How normal is grammatical development in the right hemisphere following hemispherectomy? The root infinitive stage and beyond', *Brain and language, vol. 86, no2, pp. 193-206, 2003*

Gordon, H.W. and Bogen, J.E. 'Hemispheric lateralization of singing after intracarotid sodium amylobarbitone', *Journal of Neurology, Neurosurgery, and Psychiatry*, 37, 1974

Goldberg Elkhonon, Podell Kenneth, Lovell Mark. Lateralization of Frontal Lobe Functions and Cognitive Novelty, *Journal of Neuropsychiatry Volume 6 Number 4* Fall 1994

Gott, P.S., Hughes, E.C. and Whipple, K. 'Voluntary control of two lateralized conscious states: validation by electrical and behavioural studies', *Neuropsychologia*, 22, 65-72, 1984

Gruzelier, J. and Hammond, N., 'A dominant hemisphere temporal lobe disorder?, *Journal of Psychology, Psychiatry, and Behaviour 1: 33-72, 1976*

Huerto-Delgadillo L, Anton-Tay F, Benitez-King G. 'Effects of melatonin on microtubule assembly depend on hormone concentration: role of melatonin as a calmodulin antagonist.', *J Pineal Res. Sep;17(2):55-62, 1994*

Jaynes, Julian. *The Origin of Consciousness in the Breakdown of the Bicameral Mind*, The Penguin Press 1979

Johansson Petter, Hall Lars, Sikström Sverker,1 Olsson Andreas 2. 'Failure to Detect Mismatches Between Intention and Outcome in a Simple Decision Task', *Science 7: Vol. 310. no. 5745*, pp. 116 - 119 October 2005

Joseph, Rhawn. Ph.D. *The Right Cerebral Hemisphere: Emotional Intelligence, Language, Music, Visual-Spatial Skills, Confabulation, Body-Image, Dreams, & Awareness*, Academic Press, New York, 2000

Keyes, Daniel. *The Minds of Billy Milligan*, New York: Bantam 1981

Klein, Gary. *Sources of Power: How People Make Decisions*, MIT Press, 1998

Kounios, J. Fleck, J.I, Green, D.L. Payne, L. Stevenson, J.L. Bowden, E.M. and Jung-Beeman, M. The origins of insight in resting-state brain activity, *Neuropsychologia*, 2007

Loddenkemper Tobias, Wyllie Eaine, Lardizabal David, Stanford Lisa D., Bingaman William 'Language Transfer in Patients with Rasmussen Encephalitis', *Epilepsia 44(6)*: 870-871, 2003

Lutz Antoine, Greischar Lawrence L., Rawlings Nancy B., Ricard Matthieu, and Davidson Richard J. 'Long-term meditators self-induce high-amplitude gamma synchrony during mental practice', *Proceedings of the National Academy of Sciences USA*, 2004

Matsuyama Steven S. and Lissy F. 'Jarvik Hypothesis: Microtubules, a key to Alzheimer disease(brain/tubulin/microtubule-associated protein)', *Neurobiology Vol 86, pp. 8152-8156*, October 1989

Moran Melanie and World Science staff, Vanderbilt University. *'Weird behaviour, creativity linked'*, Epub – world-science.net, 2005

Ornstein Robert. *The Right Mind*, Harcourt Brace & Company, 1997

Ostrander, S. and Schroeder, L. *Cosmic Memory*, Simon and Schuster, 1993

Pedersen, David. L. *Cameral Analysis*, Routledge, 1994

Pesenti M, Zago L, Crivello F. 'Mental calculation in a prodigy is sustained by right prefrontal and medial temporal areas', *Nature Neuroscience 4*, 2004

Ramachandran V. S. 'The Evolutionary Biology of Self-Deception, Laughter, Dreaming and Depression: Some Clues from Anosognosia', *Medical Hypotheses (1996) 47, 347-362*

Risse, G.L. and Gazzaniga, M.S. 'Well-kept secrets of the right hemisphere: A carotid amytal study of restricted memory transfer', *Neurology*, 28: 950-953, September 1978

'Savant for a Day' New York Times Magazine Date: June 22, 2003

Schiffer Fredric, MD. 'Can the Different Cerebral Hemispheres Have Distinct Personalities? Evidence and Its Implications for Theory and Treatment of PTSD and Other Disorders', *Journal of Trauma & Dissociation, Vol. 1(2)* 2000

Schiffer Frederic MD. *Of Two Minds,* Pocket Books 2000

Serafetinides EA. 'EEG lateral asymmetries in psychiatric disorders', *Biol Psychol* 19 (3-4): 237-46, 1984

Shlain, Leonard. *The Alphabet Verses The Goddess,* Penguin 1999

Srinivasan V, Pandi-Perumal SR, Cardinali DP, Poeggeler B Hardeland. 'Melatonin in Alzheimer's disease and other neurodegenerative disorders', *Behavioral and Brain Functions 2:15* May 2006

Snyder Allan, Bahramali Homayoun, Hawker Tobias, Mitchell D John. 'Savant-like numerosity skills revealed in normal people by magnetic pulses', *Perception, volume 35, number 6,* 2006

Szatkowska I, Ssymanska O, Bednarek D, Skowronska R, Grabowska A. 'Disturbances in time limited storage sensory information after right temporal lobectomy', *Acta Neurobiol Exp (Warsz)* 1996

Telfeian Albert E, Berqvist Christina, Danielak Craig, Simon Scott L., Duhaime Ann-Christine. 'Recovery of Language after Left Hemispherectomy in a Sixteen-Year-Old Girl with Late-Onset Seizures', *Pediatric Neurosurgery,* 2002

Toga, Arthur W and Thompson, Paul M. 'Mapping Brain Asymmetry', *Nature Reviews Neuroscience,* 2003

Trivers Robert. 'The Elements of a Scientific Theory of Self-Deception'

Spinney, Laura. 'I had a hunch', *New Scientist*, Vol. 159, No 2150, September 1998

Waters, Frank. *Book of the Hopi*, New York: Ballantine, 1972

CHAPTER FIVE: **Fertility and Function**

Aldridge. S. *Hair Loss: The Answers* Self Help Direct Publishing, 2000.

Clemetson, C.A.B. 'Bioflavinoids as Antioxidants for Ascorbic Acid', Symposium sui Bioflavinoidi, Stresa, 1966

Clemetson, C.A.B. et al. 'Estrogens in Food: The Almond Mystery', *International Journal of Gynaecology and Obstetrics*, Vol 15, 1978

Davies, David. *Centenarians of the Andes.* Anchor Press 1975

Domes Gregor, Heinrichs Markus, Michel Andre, Berger Christoph, and Sabine C. 'Oxytocin Improves "Mind-Reading" in Humans', *Herpertz J.Biopsych,* 2006

Exton, M.S. et al 'Cardiovascular and endocrine alterations after masturbation-induced orgasm in women', *Psychosom Med 61: 280-9,* 1999

Forsyth Alasdair John MacGregor. *'A Quantitative Exploration of Dance Drug Use: The New Pattern of Drug Use of the 1990s'* Thesis submitted for the degree of PhD, University of Glasgow, Faculty of Social Sciences, Department of Sociology, November, 1997

Gershon, Michael,. *The Second Brain*, Harper Perennial 1999.

Geschwind, Norman and Galaburda, Albert. M. *Cerebral Dominance* Harvard University Press, 1984

Gilman, Mark. 'Football And Drugs Two Cultures Clash', *The International Journal Of Drug Policy, Vol 5, No 1,* 1994

Hopkins scientists show hallucinogen in mushrooms creates universal 'mystical' experience, Epub – 11 July 2006

Kemmann, E et al. 'Amenorrhea associated with carotenemia,' *Journal of the American Medical Association*, Vol 249, No 7, 1983

Kenton, Leslie and Susannah, *Raw Energy*, Century 1984

Kripke Daniel F M.D. Kline Lawrence E D.O. Shadan Farhad F M.D., Ph.D. Dawson Arthur M.D. Poceta J S M.D. Elliott Jeffrey A Ph.D. 'Melatonin effects on luteinizing hormone in postmenopausal women: A pilot clinical trial NCT00288262' *BMC Women's Health,* 2006

Leonardi, Tom. *Ultimate Female Orgasms*, Simon and Schuster, 1995

Martini, Frederic. *Fundamentals of Anatomy and Physiology*, Simon and Schuster 1998

Motluk Alison 'The secret life of semen', *New Scientist,* August 5 2006

Murphy, Michael and White, Rhea. A. *In The Zone.* Arkana 1995

Nowak, Rachel. 'Eunachs fight back – Blocking sex hormones might help restore immunity.' *New Scientist* Vol 160, December 1998

Opar Susan. *Hallucinogens and Creativity*, Epub – lw.siena.edu

Pedersen, David. L. *Cameral Analysis*, Routledge 1994

Pierpaoli, William, Regelson, William and Coleman, Carol. *The Melatonin Miracle* New York: Simon and Schuster, 1995

Strassman, Rick. *DMT the Spirit Molecule*, Park Street Press, 2001.

Szalavitz, Maia. Gut Thoughts, *HMS Beagle, Feb 1 02 issue 119*, 2002

Szalavitz, Maia. 'The K fix', *New Scientist,* Jan 2007

Thompson MR, Callaghan PD, Hunt GE, Cornish JL, McGregor IS. 'A role for oxytocin and 5-HT(1A) receptors in the prosocial effects of 3,4 methylenedioxymethamphetamine ("ecstasy")' *Neuroscience,* 2007

'Trust in oxytocin' Editor's Summary *Nature 2 June 2005*

Wilson, Robert, Anton. *Sex and Drugs: A Journey Beyond Limits*, New Falcon Publications 1973

Wolff K, Tsapakis EM, Winstock AR, Hartley D, Holt D, Forsling ML, Aitchison KJ. 'Vasopressin and oxytocin secretion in response to the consumption of ecstasy in a clubbing population', *J Psychopharmacol,* 2006

CHAPTER SIX: **Mad Hatters and Maverick Scientists**

Baron-Cohen, Simon. 'The Essential Difference: the male and female brain', *Cambridge University, Phi Kappa Phi Forum,* 2005

Baron-Cohen Simon. 'The extreme-male-brain theory of autism', *Neurodevelopmental Disorders,* MIT Press, 1999

Baron-Cohen, Simon. The Neuropsychology of Autism and Pervasive Developmental Disorders – The Extreme Male Brain Theory, *Medscape Psychiatry & Mental Health Expert Interview,* Epub – 12/14/2005

Bishnupuri KS, Haldar C. 'Maternal transfer of melatonin alters the growth and sexual maturation of young Indian palm squirrel Funambulus pennanti', *Biol Signals Recept. 2001*

Bishnupuri KS, Haldar C. 'Profile of organ weights and plasma concentrations of melatonin, estradiol and progesterone during gestation and post-parturition periods in female Indian palm squirrel Funambulus pennanti', *Indian J Exp Biol,* 2000

Bossomaier, Terry and Snyder, Allan. 'Absolute pitch accessible to everyone by turning off part of the brain?', *Organised Sound 9(2): 181–189, 2004*

Cook CM, Persinger MA. Experimental induction of the "sensed presence" in normal subjects and an exceptional subject. *Percept Mot Skills.* 1997

Monroe, Robert A. *Far Journeys*, Doubleday 1985

Fleming Nic, Medical Correspondent. 'Autism in children '10 times higher' than first thought' *Telegraph Group Limited,* Saturday 15 July, 2006

Grimshaw, Gina M.; Bryden, M. Philip; Finegan, Jo-Anne K. 'Relations between prenatal testosterone and cerebral lateralization in children', *Neuropsychology. 9(1), Jan 1995*

Jayasundar R. 'Human brain: biochemical lateralization in normal subjects', *Neurol India* 2002

Kimata, Hajime. 'Laughter elevates the levels of breast-milk melatonin', *Journal of Psychosomatic Research,* June 2007

Knickmeyer Rebecca, Baron-Cohen Simon, Raggatt Peter, Taylor Kevin. 'Foetal testosterone, social relationships, and restricted interests in children', *Journal of Child Psychology and Psychiatry 46:2,* 2005

Lutchmaya Svetlana, Baron-Cohen Simon, Raggatt Peter. 'Foetal testosterone and eye contact in 12-month-old human infants', *Infant Behavior & Development 25,* 2002

Manning J.T, Baron-Cohen S, Wheelwright S, Sanders G. 'The 2nd to 4th Digit Ratio and Autism', *Developmental Medicine and Child Neurology,* 2001

Matsuyama S, Jarvik L. 'Hypothesis: Microtubules, a key to Alzheimer Disease', *Proc Natl. Acad. Sci. USA,* Oct 1989

Melke J, Botros LG, Chaste P, et al. 'Abnormal melatonin synthesis in autism spectrum disorders', *Molecular Psychiatry,* May 2007

Mendez M, Lim G. 'Alterations of the sense of "humanness" in right hemisphere predominant frontotemporal dementia patients', *Cogn Behav Neurol,* 2004

Persinger MA. Geophysical variables and behaviour: Ambient geomagnetic activity and experiences of 'memories': interactions with sex and implications for receptive psi experiences, *Percept Mot Skills,* June 2002

Persinger MA, Richards PM, Koren SA. Differential ratings of pleasantness following left and right hemispheric application of low energy magnetic fields that stimulate long-term potentiation, *Int J Neuroscience.* Dec 1994

Persinger MA, Tiller SG. Geophysical variables and behaviour: Increased proportions of left-sided sense of presence induced experimentally by right hemispheric application of specific (frequency-modulated) complex magnetic fields, *Percept Mot Skills* Feb 2002

Persinger MA, Tiller SG, Koren SA. Experimental stimulation of a haunt experience and elicitation of paroxymal electroencephalographic activity by transcerebral complex magnetic fields: induction of a synthetic "ghost"? *Percept Mot Skills.* Apr 2000

Randerson James. 'Too much testosterone blights social skills', *New Scientist 12 May 2004*

Roll WG, Persinger MA, Webster DL, Tiller SG, Cook CM. Neurobehavioural and neurometabolic (SPECT) correlates of paranormal information: involvement of the right hemisphere and its sensitivity to weak complex magnetic fields, *Int J Neuroscience,* Feb 2002

Snyder Allan Bahramali Homayoun Hawker Tobias D Mitchell John.'Savant-like numerosity skills revealed in normal people by magnetic pulses'
Perception, 2006, volume 35, pages 837-845

Snyder Allan, Bossomaier Terry, Mitchell John. 'Concept Formation: 'Object' Attributes Dynamically Inhibited From Conscious Awareness', *Journal of Integrative Neuroscience Vol. 3, No. 1,* 2004

Snyder Allan, Mulcahy Elaine, Taylor Janet, et al. 'Savant-Like Skills Exposed In Normal People By Suppressing The Left Fronto-Temporal Lobe', *Journal of Integrative Neuroscience, Vol 2, No. 2,* 2003

Suess LA, Persinger MA, Geophysical variables and behaviour: "Experiences" attributed to Christ and Mary at Marmora, Ontario, Canada may have been the consequences of environmental electromagnetic stimulation: implication for religious movements, *Percept Mot Skills,* Oct 2001

Tan Uner, Tan Meliha. 'Testosterone and grasp-reflex differences in human neonates', *Laterality,* 2001

Treffert Darold A, MD. 'Savant Syndrome: An Extraordinary Condition A Synopsis: Past, Present, Future', Epub – daroldtreffert.com, 2005

Treffert D. *Extraordinary People: Understanding Savant Syndrome*, Harper and Row, 1989

Serafetinides EA, Yuwiler A. 'Age, alcoholism and depression are associated with low levels of urinary melatonin.' *J Psychiatry Neurosci,* Nov 1992

Vince Gaia. 'Television watching may hasten puberty', *NewScientist.com news service,* 28 June 2004

"... So I'm dysfunctional (my experience of reality is severely limited and it is nearly impossible for me to notice). Another 'me' (that is way beyond my normal perceptual comprehension) is potentially much more functional but I need to shut down me to allow the other 'me' to take over. Sounds like religious weirdo stuff. No way can it be right. Yet amazingly the scientific evidence seems to stack up – and the key themes in ancient mythology seem to say exactly that too. But I still can't admit it's true, it would change everything that is familiar to me – no, it must be wrong and even if it was right then (according to them) it would still be well nigh impossible to take on board due to my inability to update my reality ... mmm ... intriguing ..."

Lightning Source UK Ltd.
Milton Keynes UK
07 July 2010

156678UK00005B/21/P

9 780955 678400